Tim James is the child of Welsh/English and Jamaican parents. He was raised in Nigeria, educated in England and lives in America where he is frequently mistaken for an Australian. He taught chemistry and physics for eleven years and now works full-time as an author and screenwriter. Sometimes he has a beard.

Accidental

The Greatest (Unintentional) Scientific Breakthroughs and How They Changed The World

TIM JAMES

ROBINSON

ROBINSON

First published in Great Britain in 2024 by Robinson

3 5 7 9 10 8 6 4 2

Copyright © Tim James, 2024

A CIP catalogue record for this book
is available from the British Library.

ISBN: 978-1-47214-840-7 (hardcover)
ISBN: 978-1-47214-841-4 (trade paperback)

Typeset in Scala by Hewer Text UK Ltd
Printed and bound in Great Britain by Clays Ltd, Elcograf S.p.A.

Papers used by Robinson are from
well-managed forests and other responsible sources.

Robinson
An imprint of
Little, Brown Book Group
Carmelite House
50 Victoria Embankment
London EC4Y 0DZ

An Hachette UK Company
www.hachette.co.uk

www.littlebrown.co.uk

Dedicated to Seishi Shimizu
(who loves strange stories)

Contents

PART II:
MISFORTUNES AND FAILURES

PART III:
SURPRISES

PART IV:
EUREKAS

'The most incredible thing about miracles is that they happen.'

G. K. Chesterton

It wasn't supposed to do that

....

Science is agonisingly slow. It makes progress at a rate of inches per decade, triple-checking every fact so that by the time a hypothesis is confirmed or denied, the people who originally proposed it are often retired or dead. This soul-crushing tedium might seem pointless but it's absolutely by design, because it's the best way of making sure our facts are reliable.

Things work differently in the movies, of course. Hollywood scientists always make their hasty breakthroughs at the eleventh hour from light-bulb moments, risky gambles and, more often than not, accidents.

After all, practically every superhero gets their powers by chance, e.g. they're bitten by a radioactive spider, fall into a vat of electric eels or just plain walk into a particle accelerator.* According to movies, lab accidents happen all the time – and they're always extremely useful. Now here's the really amazing thing: the movies are *not* lying. Occasionally – just occasionally – science really does work that way.

Science is supposed to be an arduous grind of shattered predictions and failed experiments, but once in a while the moons of fate align and nudge us off course to an unexpected and unintentional victory.

It's actually scary how often our species has simply lucked out. As we'll see, some of the most important life-saving inventions and profound discoveries about the Universe were only arrived at because something, somewhere, went wrong.

* This actually happened in 1978 to a man named Anatoli Bugorski who was leaning into a particle accelerator and got the full blast of a proton-beam between the eyes. Sadly he did not develop superpowers. Instead, he became paralysed on the left side of his body and suffered seizures for many, many years. (Still finished his PhD, though.)

But it's these spasms of serendipity which make science exciting. You never know when you're about to change the world, and you never know where the next big idea is going to come from. Sometimes it's not being in the right place at the right time which starts a revolution – it's being in the *wrong* place at the *wrong* time.

What counts as 'accidental'?

Writing a book about accidents really forces you to think about what the word 'accident' means and also to reflect on how useless dictionaries can be sometimes. What exactly *is* an accidental discovery, after all?

In a sense all discoveries are accidental because, by definition, we can't make them on purpose. We can't sit up one day and think 'I've decided to discover something this afternoon.' Scientific epiphanies happen when they happen, and nobody knows they're about to make a breakthrough until the moment they're making it.

That doesn't mean scientists are fumbling around depending on blind luck of course, but every step leading to a monumental revelation was taken by someone who didn't know for certain they were on the right track. They just had to hope.

Then again, if no scientific fact is arrived at on purpose, doesn't that mean *every* scientific fact is the result of an accident? Well . . . sort of.

To avoid writing a book about the literal entirety of all human knowledge, however, I decided to pin things down a little bit more precisely. I've settled on the following four categories.

PART ONE: CLUMSINESS
The purest form of accidental discovery is down to honest-to-God clumsiness, be it physical or intellectual. Here, we'll look at scientists who screwed up badly but still managed to fail their way to greatness.

PART TWO: MISFORTUNES AND FAILURES
Sometimes a mistake isn't anyone's fault, it's just rotten luck. In this section we'll look at times when everything went horribly wrong for

someone or an experiment failed to produce the desired outcome – but things still turned out OK in the end.

Part Three: Surprises

Very occasionally science is done correctly and the experiment is not a disaster. But even when everything goes right something can crop up in the results which we weren't expecting. Here we'll look at times where an incidental discovery turned out to be more important than what we were actually trying to find.

Part Four: Eurekas

These moments are extremely rare because epoch-defining ideas don't tend to pop into people's heads out of nowhere. In order for something to count as a true 'Eureka' I've defined it as: 'a major breakthrough which came to someone from a small, seemingly insignificant observation or comment'.

PART ONE

Clumsiness

··

'If you could kick the person in the pants responsible for most of your trouble, you wouldn't sit for a month.'

Theodore Roosevelt

'If you saw the mountains on my desk, nothing would surprise you!'

Albert Einstein

'D'oh!'

Homer Simpson

Boom!

The oldest record of a scientific accident comes from the Tang dynasty of ancient China (early ninth century) and consists of a written warning about a dangerous three-powder mixture that could explode without provocation.

The reader is warned not to tamper with this chemical cocktail because it was known to destroy buildings and singe men's beards,[1] but it wasn't long before people started using it anyway, mainly for fireworks and grenades.

We don't know for certain what was being described in this ancient Taoist text, but there aren't many things it could have been. There are very few three-way reactions in chemistry, and even fewer that lead to explosions. It's therefore a reasonable assumption that this is the very first recorded reference to gunpowder.

Tradition asserts that the monks who made it were trying to discover the elixir of life, but it's far more likely they were simply trying to make fertiliser. Gunpowder is composed of powdered charcoal, sulfur and saltpetre. The latter two components are important plant nutrients, so the early Chinese botanists were probably mixing sulfur and saltpetre together for a good crop yield and somehow got it mixed up with charcoal.

When you heat this mixture, the three powders react and their molecules rearrange to make nitrogen and carbon dioxide. This rapid production of gas creates a devastating shock wave as the surrounding air is pushed aside in order to make room. In other words, it detonates.

Whether the monks were trying to live forever or just get healthier plants, they ended up making the first high-performance explosive and gunpowder became the go-to projectile weapon fuel for centuries. Until, that is, a German chemist named Christian Schönbein had a disaster of his own which improved the recipe.

Cottoning on

Schönbein was a keen and well-respected scientist who had already discovered ozone and invented the fuel cell, but his wife did not appreciate experiments in the home and forbade him from carrying them out. However, during one particular afternoon in 1845 while she was out, Schönbein did what anyone would do when he thinks he's got the place to himself – he snuck into the kitchen for a spot of clandestine chemistry!

Whatever experiment he was *meant* to be doing is unknown, because while preparing his reaction Schönbein knocked two large beakers onto the table, one containing sulfuric acid, the other containing nitric.

Panicking at the danger (and no doubt the prospect of having to explain all the acid damage to his wife), Schönbein grabbed her cooking apron and began soaking up the corrosive brew as fast as he could. Once he had got the majority of it up, he transferred the apron to the oven in the hope of getting it dry, but when he did so things got worse. Specifically, the oven exploded.

Schönbein didn't know the explanation for what was going on, but we can explain it today. Cotton is made of a polymer called cellulose which, when heated with nitric acid, reacts and incorporates molecules of the acid into its structure. You need a bit of sulfuric acid to get the reaction going and when you do, the result is an extremely combustible fabric called nitrocellulose.

Schönbein had combined the cellulose in the apron with the nitric acid he was trying to mop up and just so happened to provide the perfect sulfuric acid catalyst to get them reacting. All it needed was a bit of heat to ignite, which he introduced in the form of the oven. He had turned his wife's apron into guncotton.[2]

Gunpowder had been the standard explosive for two millennia, but it had several drawbacks. Firstly, it created a thick smoky discharge which made battlefields impossible to navigate once the cannons started firing. Secondly, it took a lot of heat to detonate. And thirdly, as soon as gunpowder gets even a little bit damp it stops working and the only way to dry it out is to heat it . . . (pro tip: heating gunpowder isn't something you want to do).

Schönbein's guncotton, on the other hand, burned without much smoke, ignited without much heat and could get damp while retaining potency. As an added bonus, guncotton produces five times more gas than gunpowder, i.e. it gives off five times the explosive force. It quickly took over from gunpowder and became the standard blasting stock for weapons. But that wasn't all; nitrocellulose had another chemical gift to give us.

SHATTERED

In 1903 the French chemist Edouard Benedictus was pottering around his lab when he knocked one of his glass beakers off the shelf. Instead of shattering to pieces when it hit the floor, the glass stayed intact. Benedictus had been making nitrocellulose in the beaker the previous day, and realised he hadn't done a good job of cleaning it. There was still a thin film of nitrocellulose lacquer stuck to the inside. This filmy version of the chemical was transparent, sticky and apparently very strong.

Initially, Benedictus thought nothing of it, but a few years later, while reading a newspaper article about car crashes, he realised the implications. The article described how large numbers of people were injured in car accidents not from the impact but from flying shards of glass. Benedictus remembered his shatter-proof beaker and set to work immediately.

He spent the next twenty-four hours working non-stop to perfect his invention and ended up with something that solved the problem. By sandwiching a sticky sheet of nitrocellulose between two sheets of glass, he had a material that was completely transparent but would not shatter when struck (since the nitrocellulose film held the glass shards in place).

He marketed his invention as Triplex™ and found its first major use in eyepieces for gas masks before it was later incorporated into windscreens, windows, television screens and eventually bulletproof glass. Nitrocellulose not only gave us the technology to create efficient projectile weapons but also the technology to shield ourselves from them.[3,4]

Action

The next big use for nitrocellulose was discovered in 1855 by the British chemist Alexander Parkes, who was studying the widely used commercial material shellac.

Shellac is a sticky resin excreted by female Indian lac bugs as they stick their eggs to tree branches for protection. In the nineteenth century it was widely used for mouldings and casings, but Parkes wanted to find an alternative to making everything out of sticky insect egg paste because . . . well . . . wouldn't you?

He tried mixing a number of naturally occurring polymers in the hope of creating a rigid, lightweight material and one evening decided to mix nitrocellulose with camphor wax, dissolving the solution in alcohol. He heated it, hoping it would harden into a resin, but the alcohol evaporated, leaving a flexible rubbery lump in the bottom of the flask. Not what he had hoped for. What he had just made, however, was one of the most important chemicals in history: celluloid, the very first synthetic plastic.[5]

Parkes never made any money out of his flexible new substance because he couldn't find uses for it. It could be used to make billiard balls when cooked in spherical moulds, but beyond that it had little application. It was another thirty years before the French inventor Louis Le Prince used celluloid to make the 35mm film used in motion picture cameras, giving us the movie industry.

Celluloid is not as combustible as its parent chemical nitrocellulose, which is fortunate because film projectors get very hot, but it can still degrade over time and even self-ignite in a heated room (as one too many film archivists have discovered). In fact, celluloid fires (like the one Quentin Tarantino used for the finale of his film *Inglourious*

Basterds) are almost impossible to extinguish because when celluloid burns it produces its own oxygen and keeps the fire self-perpetuating, even underwater.[6]

Fortunately, celluloid's tendency to spontaneously combust is rare, and most flammable materials need a source of ignition to get going. But how do we get a fire started in the first place? For that we turn to yet another accidental invention.

STRIKE A LIGHT

In 1826 the English chemist John Walker was experimenting in his home to find a new fuel source. His approach was the highly sophisticated technique of stirring flammable chemicals together, gumming them to the end of a stick and holding them over the fire in his hearth to see what would ignite.

One evening, Walker was trying out a new recipe, but as he moved his stick through the air, he scraped it against the brick hearth of his fireplace and it burst into flame. Walker had just invented the matchstick.

He never patented his formula, however, and a number of different variations were tried by other inventors across Europe, all working on the same principle. The trick is getting a fuel molecule to vibrate fast enough into an oxygen molecule so the two will rearrange into stable products, releasing heat in the process.

The most effective recipe, the one most often used in matches today, consists of a chemical called phosphorus sesquisulfide caked together with another called potassium chlorate.

Phosphorus sesquisulfide acts as a fuel, because phosphorus and sulfur atoms bond to each other weakly, i.e. given a chance they would much rather bond to oxygen. The other chemical, potassium chlorate, contains three oxygen atoms, making it a rich oxygen source. It's also very unstable, but in the opposite sense – the oxygen atoms aren't well bonded to each other and would be happier bonding to phosphorus and sulfur.

If you pack these chemicals next to each other you've got an unstable blend that's aching to react, and if you scrape them against something rough, e.g. the bricks of a fireplace, the powdered glass of a matchbox or Clint Eastwood's face, you start vibrating the molecules against each other with enough energy to start them swapping oxygen and *voilà* – a fire!

Safety matches do the same thing except that the chemicals are separated, with the head of the match containing potassium chlorate (source of oxygen) and the surface of the matchbox containing phosphorous sesquisulfide (desperate for oxygen).

For the outstanding – albeit unintentional – achievement of inventing the matchstick, John Walker was immortalised as a statue in his home town of Stockton-on-Tees – until 1990 when it was discovered the statue everyone had been admiring was modelled on the wrong John Walker. By complete accident, the sculptor had mixed up John Walker the chemist with an actor from London also named John Walker.[7] A fitting tribute, perhaps.

BOUNCING BACK

Charles Goodyear was born in Connecticut in 1800 and spent the first thirty-nine years of his life as a failed businessman. His father, Amasa Goodyear, was an entrepreneur who manufactured the first pearl buttons, but Charles did not inherit his father's nose for industry. After initially considering a career in the church, at the age of seventeen he decided to follow his father into manufacturing and began numerous enterprises with patented designs for shoes and furniture. Sadly all his endeavours flopped and he frequently found himself in and out of jail for debt. Nevertheless, he was convinced God had chosen him to invent something which would change not only his fortunes, but the world.[8] He was right.

In 1834, while staying in New York city, he encountered a wondrous material called latex. Latex is produced by trees common to tropical regions and can be extracted as a white gloopy liquid. When left to dry, it coagulates into a bouncy lump which the English chemist Joseph Priestley (discoverer of oxygen) used to 'rub out' mistakes in his note-book. The men who sold him the latex started calling the substance 'rubber'[9] – after Priestley's use for it – and the name caught on.

The reason for rubber's bounciness and flexibility is down to how the atoms are arranged. Every atom has preferred angles at which it will bond to others, determined mostly by the shapes of the atoms themselves. For instance, oxygen tends to bond to other atoms at a 104.5-degree angle, while carbon prefers to bond at 109.5 degrees, and so on. Any deviations from these 'ideal angles' are a strain and the atoms will twist to get out of them.

Rubber is made from chains of atoms arranged in zigzags, but when these chains are pulled tight, the atoms are forced into a strained line.

This isn't their preferred configuration and as soon as they are released, they'll snap back like tiny springs in order to get stable, making rubber highly elastic.

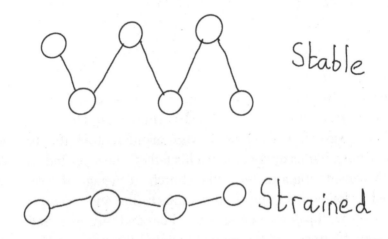

Stable

Strained

(Speaking of springs: the engineer Richard James invented the Slinky™ while trying to build a ship stabiliser. He was testing different thicknesses of steel spring when he knocked one of his prototypes off a shelf. It did the characteristic backflip down to a pile of books, before carrying on to his workbench and finally the floor. The kids in his neighbourhood were so excited by his 'walking spring' that James decided to sell it as a toy, with his wife suggesting the name Slinky™ since it sounded like something sleek and elegant.[10])

Back to Charles Goodyear, who saw potential in rubber as a flexible super-material. He wanted to use it for his various patents but in order to do so he would have to overcome its main problem – extreme sensitivity to temperature.

When rubber gets warm it starts melting and becomes a sticky mess as the atoms inside the chains break their bonds. This happens at around 37°C which is, unfortunately, human body temperature. Meanwhile, at the other end of the spectrum, when things get cold the atoms become rigidly held in place and refuse to budge, making the rubber brittle. Rubber melts on a warm day and shatters on a cold one.

One shoe manufacturer Goodyear spoke to told him that his company was close to collapse after it lost $20,000 in melted shoes which were not only unsaleable, they had to be buried in the company's landfill because the smell was so obnoxious. By this time, of course, Goodyear was no stranger to financial risk. He knew that whoever could make rubber a practical material would change the shoe industry, so he set about experimenting with the substance to see if he could alter it.

For five years he tried all sorts of experiments, without success (sometimes within a debtor's cell). He tried mixing rubber with powdered bronze, nitric acid, lead, magnesium and magnesium carbonate – at times producing dangerous gases that nearly suffocated him. On one occasion he even made a pair of rubber trousers which melted to his assistant's legs, welding him to a chair.

During these years, Goodyear gained a reputation as a rubber-obsessed eccentric who could be seen walking the streets of New York in a hat, shoes, cape and gloves all made from rubber, while carrying a rubber newspaper, none of which was flexible at anything but room temperature.

Many would have given up after these failures but Goodyear believed, rather presciently, 'that which is hidden and unknown, and cannot be discovered by scientific research, will most likely be discovered by accident'.

Which is precisely what happened. One evening in the winter of 1839, Goodyear's dogged belief that he would accidentally solve the rubber problem was proved correct. He had recently made the acquaintance of another rubber-obsessed maniac named Nathaniel Hayward who had been shown in a dream that mixing rubber with sulfur would improve its heat resistance. Goodyear liked the idea and, putting up with yet another stench, began incorporating sulfur into his rubber samples.

It was while enthusiastically showing one of these sulfur-rubber lumps to his brother in his kitchen that Goodyear accidentally let go of the sample while making an excited gesture. It subsequently flew across the room and landed on the stove. Everyone knew this meant bad news because rubber always melted in heat, but as Goodyear peered into the pan he saw something astonishing. The rubber was not only remaining solid, it was also forming a tough sheen similar to hardened leather. Heating rubber on its own will destroy it, but baking the hell out of it with sulfur apparently had the opposite effect. The rubber was surviving.

Sulfur atoms bond very well to the atoms in rubber chains, forming cross-links between them. The chains themselves are unchanged by this process, meaning the material still has its flexible characteristics, but the chains are now firmly held together by sulfur.

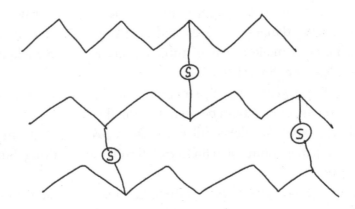

This means that at high temperature the chains won't separate, while at low temperature the structure will not become brittle. By dropping the rubber into the pan, Goodyear had discovered the perfect method to reinforce it.

He nailed the sample to the front door of his house and, when he examined it in the morning, found it had survived the harsh cold of a New York winter's night. He had finally discovered the secret to making durable rubber and called the process 'vulcanisation' after Vulcan, the Roman god of fire.

At first his technique was met with intense scepticism by his friends and the rubber industry; this wasn't the first time Goodyear had claimed to have made a discovery which turned out to be nothing. But in time, people began to realise that Goodyear rubber really was the holy grail of elasticity.

Sadly, a number of patent disputes and legal battles over the origin of vulcanisation used up all the money Goodyear made from his invention and he died bankrupt.[11] A relentless scientist he may have been, but a shrewd businessman he was not.

Nonetheless his method lives on and, as you will no doubt have guessed from his surname, became invaluable to the car industry. Put simply: vehicles would not have wheels without Charles Goodyear and his slip of the hand. Speaking of which . . .

EMPLOYEES MUST WASH HANDS

In 1938 the Swiss chemist Albert Hoffman was researching Portuguese ergot fungus. Ergot was known to treat migraines and labour pains, but Hoffman was hoping to test its use as a respiratory medicine. One of the chemicals he extracted was lysergic acid which he decided to convert into other chemicals by globbing stuff onto the molecule at random. This might sound haphazard, but it's a standard approach for biological chemistry. You start with one molecule and try adding bits to see whether it improves the function or not.

At some point during this process, Hoffman manufactured a chemical called LySergic acid Diethylamide-25 (LSD for short). It didn't have many obvious applications and so he forgot about it for five years. But in 1943, Hoffman decided to revisit it and see if he had missed something. On 16 April he re-synthesised the compound, but in the process, started to feel dizzy and had to abandon his work.

After barely making it home, Hoffman went to bed and suffered the most vivid nightmares of his life. Wondering what he had eaten to make himself so sick, he headed to work the following Monday and noticed the reaction flask he had been using. Had he got some of the residue on his skin? Was that the cause of his sickness?

In order to see if it had been the LSD which caused the sickness and dreams, Hoffman could think of only one thing to do: deliberately swallow a grain of pure LSD and see what happened. Once again, he began feeling extremely sick within an hour and had to leave his post, this time with the aid of an assistant.

They made their way back to Hoffman's house on bicycles, while Hoffman started having what he called a 'severe crisis' in which he hallucinated that people around him were turning into demons. He was having the world's first acid trip.

After asking his neighbour to bring him some milk (before accusing her of being a witch) Hoffman lay on his bed for several hours as the hallucinations got worse. Starting to get nervous he had done himself permanent damage, he called for a doctor who found nothing physically wrong with him and concluded that he hadn't poisoned himself in the conventional sense. He had instead discovered a psychoactive chemical which caused extreme fracturing of sensory perception.[12] LSD was a potent hallucinogenic.

Nobody knows for sure how LSD works because nobody knows how perception works. What *is* known, however, is that LSD triggers the excessive release of a neurotransmitter called glutamate. Glutamate has a number of functions but it is chiefly involved in the formation of memories. It's possible that LSD leads to a glutamate overdose which causes the brain to 'over-remember', i.e. bring up disconnected memories that have nothing to do with what the user is experiencing. This confusion between remembering unconnected things combined with actual sensory information may lead to hallucination.[13]

Hoffman's employer, Sandoz laboratories, initially marketed LSD as a medicine to cure alcoholism and sexual perversion, but in 1950 the US Central Intelligence Agency (CIA) purchased the world's supply of it and began using it to experiment on people as part of the now infamous MKUltra project.[14]

TIP OF THE TONGUE

A less controversial and far more widely loved chemical was discovered in a similar way. In 1879 the Russian chemist Constantin Fahlberg was working in the chemistry lab of Johns Hopkins University, studying coal tar – the sticky black substance you get after you burn coal. After cleaning up for the day, Fahlberg went home and found his bread tasted outrageously sugary. As did the water he drank and the napkin with which he dried his moustache.

He knew it wasn't the food in his home so it had to be something on his fingers he had picked up from the lab. He went back immediately and began tasting all the different chemicals he had extracted from the coal to find out which one was the sweetest. In his own words: 'I dropped my dinner and ran back to the laboratory. There, in my excitement, I tasted the contents of every beaker and evaporating dish on the table. Luckily for me, none contained any corrosive or poisonous material.'[15]

Eating all the chemicals in a laboratory is discouraged as a rule, although a former lecturer of mine (who shall remain nameless) used to make a habit of deliberately consuming micro-quantities of everything he ordered from the chemical supplier, insisting that keeping the dosages below the toxic threshold made it safe. A slightly worrying behaviour, but he's the only person I've met who can describe what cyanide tastes like.

Eventually, after much trial and error, Fahlberg identified the culprit that was sweetening his food: ortho-benzoic sulfimide, the first artificial sweetener, three hundred times sweeter than table sugar.

Fahlberg sought patents for his chemical and renamed it saccharin after the word 'saccharine' but it was soon investigated as a harmful substance, being especially demonised by one of the US Food and Drug

Administration's (FDA) senior researchers, Harvey Wiley, who considered it an unnatural and unhealthy sugar substitute.

The FDA's investigation was blocked, however, by none other than President Theodore Roosevelt, who had recently been prescribed saccharin as a sugar alternative by his physician Presley Rixey in order to help lose weight. You can't mess with the president's weight-loss programme, so the FDA's attempts to ban saccharin were thwarted, with Roosevelt saying directly to Wiley's face: 'Anybody who says saccharin is injurious to health is an idiot.'[16]

However, Harvey Wiley eventually got his way in 1911 when Roosevelt was out of the White House and he managed to get saccharin classified as 'an adulterant'. This decision was overturned a few years later during the First World War, however, when the US Government decided they needed saccharin in soldiers' food because sugar was in short supply.[17]

Sweet tooth

Curiously, this isn't the only time a sweetener has been discovered by someone accidentally tasting it on their hands. In 1937 Michael Sveda at the University of Illinois was researching anti-fever medication when he decided to take a cigarette break. He put his cigarette down on the desk and when he put it to his mouth, discovered an alarming sweetness on the tongue.[18] He had accidentally dipped his cigarette in sodium cyclohexylsulfamate, known today as cyclamate – forty times sweeter than sucrose and the chief sweetener in the American sugar substitute Sweet'n Low™.

Then again in 1965 the chemist James Schlatter was researching anti-ulcer drugs when he got chemical residue on his fingertips, licked them in order to turn over a piece of paper, and found his fingers tasting sweet.[19] He had discovered aspartame, two hundred times sweeter than sucrose and used in Diet Coke™.

And then, yet *again*, to the point where you start to think someone up there is playing a prank, in 1976 the chemist Shashikant Phadnis was working on chlorinated sucrose when his supervisor Leslie Hough asked him to test it. Phadnis, who was from India and didn't speak English as a first language, misheard and thought his supervisor was instructing him to 'taste it'. Like any loyal grad student, Phadnis blindly obeyed and discovered sucralose[20] – six hundred times sweeter than sucrose, now marketed as *Splenda*™.

Saved by A. Bell

It's not just physical clumsiness which can move human innovation forward. Sometimes it's a clumsy translation – like the one which inspired a device you've probably used a dozen times today already.

Alexander Graham Bell was born in 1847 to A. M. Bell, a respected Scottish elocutionist and his wife Eliza, who had been deaf since the age of twelve. Alexander, known as Aleck to his family, was originally christened Alexander Bell after his father, but was given the middle name Graham as a birthday present when he turned eleven. Efficient parenting right there.

From a young age, Aleck was interested in the scientific study of speech. He grew up in a community of linguists, elocutionists and speech therapists, including Napoleon Bonaparte's nephew Lucien who was known for having translated *Song of Songs* (the Bible book with lots of sexy bits) into over twenty languages.

As a child, Aleck devised a communication method with his mother that involved humming into her forehead to signify letters and he even taught the family dog to manipulate its mouth to mimic words.

By the time he was twenty and married to Mabel Hubbard (who was born deaf) Aleck was conducting rigorous experiments into oral acoustics. He would take pencils and hold them to his lips, throat and nose, testing to see where vibrations were coming from and how the throat produced the sounds of language.

Eager about his discoveries and research, he wrote a forty-page summary which he sent to his father, wondering if they would be worth publishing. His father was excited by his son's discoveries and showed them to his friend Alexander Ellis, who unfortunately had deflating news. The German physicist Hermann von Helmholtz had already been conducting similar experiments and had written a whole book on the subject. Aleck was too late.

Helmholtz had carried out one particularly impressive experiment in which he got tuning forks to mimic vowel sounds. By positioning three forks in front of each other to represent the front, middle and back of the mouth, Helmholtz rigged each fork to vibrate with a battery. When he got the right frequencies and the right distances between the forks, Helmholtz was able to artificially create synthetic vowels in the air.[21]

A. M. Bell was a little sad that his son's work had already been poached but posted him a copy of Helmholtz's book anyway, hoping it would show his son that at least he was on to something. Aleck gratefully received the tome, but there was a problem. The book was in German and he didn't understand any of it. Exactly why his father sent his son a book in a language he couldn't read is unclear. Aleck used to perform a party trick in which he could look at the words of a foreign language and accurately deduce how to pronounce them, so perhaps his father had forgotten his son didn't *actually* speak German and could only imitate its sounds.

Either way, Aleck pored through the book and tried to figure out what was going on from illustrations and rough attempts at translation. He didn't know anything about electricity or German, though, and when he looked at the diagrams of tuning forks and circuits, he misunderstood them completely.

He thought Helmholtz had devised a way to transmit sounds electronically down a wire from a receiver to a transmitter, and decided that since Helmholtz had shown it was possible, creating a long-distance sound machine should be his next goal. So he set about teaching himself the physics of electricity.[22]

By 1875, Bell had become a skilled engineer and had perfected a device which would transmit the full range of human speech electronically – what we now call a telephone. It's a device so brilliantly simple that its design has not changed much since Aleck's original.

When someone speaks into the receiving end, a disc of paper (or nowadays plastic) vibrates in time with the air waves being generated.

This paper diaphragm is attached to a magnet, so that as the diaphragm vibrates, the magnet does too. Bell had learned that when you vibrate a magnet inside a coil of wire it generates an electrical disturbance which can travel down the wire.

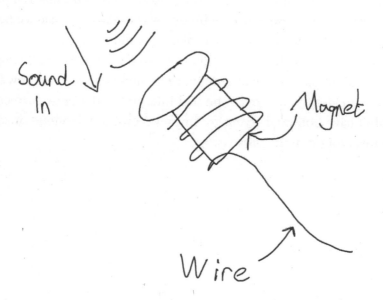

At the other end of the wire, a second magnet can receive this electrical signal and vibrate in time with it. This magnet is also attached to a diaphragm which will generate vibrations in the air identical to those at the source. The microphone turns vibrations into electrical signals, while the speaker reverses the process.

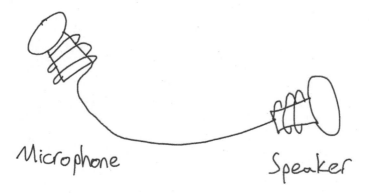

Bell believed he was simply building a practical modification based on Helmholtz's earlier work – but Bell had misunderstood the book and ended up inventing something entirely new. The notion of sending continuous sound signals down wires had been dismissed by the engineering community as impossible. You could send blips and pulses via Morse code, but continuous vibration was the work of science fiction. But Bell didn't know it was impossible, so he did it anyway.

When he eventually learned his mistake (and finally read a copy of Helmholtz's book in French, a language he spoke) he realised how fortunate the accident had been. If he had interpreted the book correctly he wouldn't have bothered trying to 'improve' on the telephone, inadvertently inventing it in the process.[23]

MISSING THE BELL

The charming story of Alexander Graham Bell making a breakthrough because he thought it had already been done is not the only one of its kind.

In 1939, at the University of California, Berkeley, a twenty-five-year-old mathematics student named George Dantzig arrived late to a lecture being given by his supervisor Jerzy Neyman. Dantzig copied down the homework problem from the blackboard, handed in his solution a few days later and thought nothing of it.

What Dantzig hadn't realised was that the question on the board was not a homework assignment. It was one of the most difficult, unsolved mathematics problems in history and he had just solved it. Had he been on time to the lecture he might never have given it a shot.[24]

If there is a moral in both these stories it's that you should never warn people how difficult something is going to be because it will prime the prophecy to fulfil itself. Perhaps the best thing to do when tackling something near impossible is to simply assume it's already been done.

KEEPING THE BEAT

In the early 1950s Wilson Greatbatch was studying for a degree in electronics from Cornell University in New York State and working at a local lab to make ends meet. While there, he encountered surgeons experimenting with 'bioelectricity' – electrical circuits in the body which regulate organ function, especially those of the heart.

A heart is a muscle which twitches twice per second when a shock of sixty-thousandths of a volt is zapped across it. If the nerve in charge of delivering this signal fails (as sometimes happens in old age) the heart doesn't get its stimulation and beats irregularly, not enough, or not at all.

The only solution for keeping people's hearts beating in these circumstances was a device consisting of a large metallic ring sewn under the skin, connected to the heart through wires. Outside the skin, a magnetised plate was attracted to the ring through the flesh, which was then connected to a battery the size of a lunchbox the patient had to lug around, with a dial to increase or decrease heart rate as needed.[25]

Greatbatch was impressed by the ingenuity of these devices but thought the design was impractical and clunky, not to mention a risk. The battery and metal ring were separated by a small layer of chest skin and anything could knock it out of place.

By 1958, Greatbatch (now a lecturer of electronic engineering at the University of Buffalo) was playing with electrical devices in his shed to try and build something to record heartbeats. One of the crucial components he was using for his device was something called an oscillator circuit.

No bigger than a matchbox, an oscillator circuit has the job of converting DC electricity into AC electricity. DC is when all the electrons in a wire flow in a continuous loop round the circuit, while AC is when electrons shuffle back and forth on the spot (the electricity from a UK plug socket is doing this shuffle fifty-five times per second). An oscillator's job is to take a smooth DC current from a battery and turn it into a rapid-fire AC shuffle.

As he was working on his circuit, Greatbatch reached into his toolbox to pull out a resistor – a device which reduces electrical current. They're about the size of a grain of rice and have coloured bands telling you how much current they will resist. But on this occasion Greatbatch misread the coloured bands and plucked the wrong size.

His oscillator circuit needed a ten-thousand-ohm resistor, but he had accidentally put in a ten-million-ohm resistor instead. With this hugely obstructing component, the circuit could no longer snap the electrons back and forth rapidly. Instead, he got what he described as a 'squeg' effect: a sharp electrical pulse occurring about once every second. Essentially, the overly obstructive resistor had turned a rapid-fire shuffle into a much more sluggish drumbeat. Greatbatch realised what he'd done at once: he'd invented a circuit small enough to fit inside the body which could deliver electrical pulses to a rhythm. He had invented the pacemaker.

He set to work perfecting his design and spent his life savings (as well as his wife Eleanor's) converting his shed into a pacemaker manufacture facility.

At first, his pacemaker was not popular among surgeons who doubted it could work, so Greatbatch had to find a way to prove it. He decided to perform some home surgery on his pet dog and implanted one of his pacemakers in its chest, where it worked perfectly (although this first attempt eventually stopped working after a few hours because he wrapped the pacemaker in electrical tape which wasn't waterproof).[26]

Once he had proved his pacemaker capable of delivering the required shocks inside a living being, the medical community acknowledged his design. Greatbatch's pacemaker business was valued at $50 million by the time he died and over six hundred thousand people are fitted with them every year.

ON THE USES OF MOULD

Having a slobby desk ended up leading to one of the most important biological discoveries of the twentieth century, perhaps coming second only to the solving of DNA (see Part Four for how luck helped *that* one along).

The story of Alexander Fleming's desk and how it changed the world is legendary, and there's a good chance you already know it. But there are more accidental moments in the story of antibiotics than people realise. Perhaps too many.

The Scottish physician Alexander Fleming had a fascination with biological nasties. He was so enamoured with fungi and bacteria that he used to paint with them, decorating agar canvases with moulds to create images which he showed to everyone who visited his lab, including King George V and Queen Mary (who were rather puzzled by the point of them).[27] But on three separate occasions, one deliberate, one partially deliberate and one not deliberate at all, Fleming made major discoveries that pushed our ability to fight infection over the horizon.

Fleming first became interested in bacterial research during the First World War when he treated wounded patients along with his mentor Almroth Wright. Wright and Fleming were researching in field hospitals and discovered that when a bullet punctured a soldier's clothing it was actually bacteria from the dirty uniform getting into the wound which led to the most serious infections.

Together they discovered that the standard antiseptics used to treat wounds were only killing bacteria on the surface of the skin and not getting deep enough to prevent sepsis. Worse, Fleming discovered that the most commonly used antiseptic, flavine, was actually killing cells of the immune system. By administering too much antiseptic, doctors could be making infections worse.

By the time he was able to convince other medics, the war was drawing to a close. It's not known how many lives were lost due to improper use of antiseptics.[28]

His second big discovery happened in 1921. This time, he was growing moulds in Petri dishes while he had a cold. According to the official account he gave to the Royal Society he decided to deliberately test his own nasal discharge to see what it contained, but another account states his nose was simply running and he let it drip into the Petri dish before waiting to see what would happen.[29]

To his delighted surprise, his sputum contained a hitherto undiscovered enzyme called lysozyme which had antibacterial properties. It later turned out that lysozyme is one of the body's natural antibacterials and can be extracted from other readily available (and less gross) fluids like egg whites.

But then came the mother lode discovery. In August 1928, Fleming went on holiday with his family and left his lab in a sorry state with Petri dishes piled in the sink. They should have gone into an incubator but Fleming didn't bother and left them unattended for weeks. When he returned on 3 September, he went to his lab and discovered a number of moulds had grown in the dishes.

He began chucking them one by one into the cleaning bath which would corrode the mould, but as he came to the end of the pile, he picked up a dish which looked odd. He was about to chuck it into the bath when he stopped and saw that the mould had destroyed the bacteria around it. Usually, bacteria in a Petri dish will clump to a mould and overtake it, but this time it hadn't. There was, instead, a clear boundary all the way around the outside of the mould circumference.

Fleming looked at it and said aloud: 'That's funny.' It appeared as though the mould contained something lethal to bacteria. His lab assistant Daniel Pryce reminded him that a similar moment had led to the discovery of lysozyme, so maybe he had discovered another antibacterial? Fleming had in fact discovered the *ultimate* antibacterial.

In a lab downstairs, the mycologist Charles La Touche had been working with a mould called *Penicillium rubens* and some of the mould spores had got caught on the wind, made their way up the stairs and landed in Fleming's Petri dish while he had been on holiday. There, it had grown into a stable colony and started secreting antibiotics.

We sometimes think of all micro-organisms as our enemies but bacteria, fungi and viruses are not out to get *us* specifically, they're at war with each other too. *Penicillium rubens* manufactures its own antibiotic which Fleming managed to extract, originally calling it 'mould juice', before settling on the name by which it is now known: penicillin.[30,31] The age of antibiotics had begun.

The mechanism of penicillin is thought to be due to an unstable four-atom ring structure in the centre of the molecule. The atoms in this ring are so tightly squashed that the ring is looking for an excuse to pop open. As soon as penicillin molecules encounter the wall of a bacterium, they find the right kinds of atoms to bind to and cleave themselves, disrupting the membrane of the bacteria and forcing it to form holes which bursts the bacterial cell.

Fleming was later awarded the Nobel Prize and, deliberate or not, his discovery changed the way medicine was done. Most of the antibiotics we use today originated in soil samples collected everywhere from farm-yard potting plants (streptomycin in 1943)[32] to the jungles of Borneo (vancomycin in 1953).[33] The rule seems to be that, wherever there is mould, there are antibiotics.

Using your melon

It wasn't just soils that yielded the best moulds. In 1943 the American biochemists Robert Coghill and Ken Raper were asked to find a new type of antibiotic to help wounded soldiers as part of the Second World War effort. Penicillin was great at its job, but it wasn't easy to mass-produce – so Coghill and Raper were charged with finding a mould which could produce something as good as penicillin but which would be easier to grow.

They had moulds sent to them from all over America for testing, but the one which turned out to be the best came from the bacteria *Penicillium rubrum* which was scraped off a mouldy cantaloupe bought from an Illinois farmers' market.

The identity of the scientist who bought the melon and scraped the mould off is lost to history, although she was given the nickname 'Mouldy Mary', suggesting it may have been a technician who worked in the lab called Mary Hunt.[34]

DISINFECTANT TO THE RESCUE

Penicillin was initially administered through injection, but another bizarre twist helped us along yet again. In 1945 the French chemist Michel Rambaud decided to set up a penicillin factory in a disused brewery with his business partner Richard Brunner.

Unfortunately, the tanks where he was trying to grow penicillin kept getting contaminated with *E. coli*, so he decided to clean them with phenoxyacetic acid – a standard disinfectant. But to Brunner and Rambaud's surprise, this somehow led to a blossoming *increase* in penicillin production.[35]

Unbeknown to them, the mould which produces penicillin actually feeds on phenoxyacetic acid as a nutrient and uses it to make an antibiotic called phenoxymethylpenicillin. This new antibiotic was resistant to acid, meaning it could survive the human stomach and be taken orally.

So the discovery of antibiotics, the mass-produced version of them and the ability to take them orally, are all down to flukes. Nature isn't always trying to kill us.

SOMETIMES IT IS, THOUGH . . .

Peptic ulcers are small rips in the stomach lining. When you get one, acid leaks through to the surrounding organs and starts corroding whatever it touches, which is as bad for you as it sounds. Peptic ulcers are the number one cause of stomach cancer so if you don't treat them early, you're in serious danger.

In the 1970s everyone believed peptic ulcers were caused by stress and the best way to treat them was to calm the patient down with anxiolytics (anti-anxiety medications). But an accidental encounter led Australian researchers Barry Marshall and Robin Warren in a different direction.

A patient they were treating had a vicious ulcer in his stomach, but at the age of eighty, there wasn't much they could do for him. The anxiolytics they'd typically prescribe hadn't been tested on people that old so it wouldn't have been ethical to administer them. The only option was to give him antibiotics which would do nothing for the ulcer but would at least stop it getting infected.

A week later, the man returned with a spring in his step and a smile on his face – his ulcer was gone. The question now facing Marshall and Warren was how a course of stop-gap antibiotics had treated a stress-related illness?

Nobody had ever gone looking inside stomach tissues for bacteria because it would be like looking for bears at the bottom of the ocean – it's so obvious they can't be there nobody even thinks to try. But Marshall and Warren wanted to know what was going on, so they began analysing stomach tissue from ulcer patients and found a strain of bacteria so tough it could survive the intensity of digestive acid: *Helicobacter pylori*. Was it possible that stomach ulcers, one of the most dangerous maladies known, were the result of a simple bacterial infection?

In order to find out whether this bacteria was responsible for the ulcers, Marshall and Warren would have to show there were higher amounts of it in the stomachs of ulcer patients.

A small sample of bacteria can be hard to spot under a microscope, so the best way to examine someone's gut biome is to extract digestive fluid and smear it on a Petri dish with food. Then, you leave it for forty-eight hours during which time the bacteria will multiply and spread across the dish.

Warren and Marshall went through thirty-three patients with ulcers, finding no trace of *Helicobacter pylori* in any of their stomach juices. They were close to giving up, but on the thirty-fourth patient they made a mistake. They accidentally left the Petri dish in the lab over the Easter holiday weekend, forgetting to throw it out. It sat there for four days and when they came to examine it on the following Tuesday they found it coated with *Helicobacter pylori*.

It turns out that *Helicobacter pylori* grows at a slower rate than normal bacteria, so the usual forty-eight hours used as a standard test had not been long enough for it to develop. They had probably gathered thirty-three samples successfully and just not noticed. But now they had done it – proof that ulcers were treatable with antibiotics.[36]

They published their results but everyone dismissed them because what they were suggesting was so loopy. They would have to work harder.

They tried infecting mice, rats and pigs with the bacteria but had no success since *Helicobacter pylori* turns out to be harmless to those species. By 1984 they were running out of options and starting to sound like cranks, so Marshall decided there was one more thing to try. Give himself an ulcer.

After collecting stomach juices from a patient with severe ulcers, Marshall assumed he would get a mild case if his hypothesis was correct. One evening, without telling anyone, he got a vial of recently obtained stomach juice and necked it. Two days later, he woke in the middle of the night with an agonising pain in his abdomen.

A quick endoscopy revealed that he had successfully given himself a powerful dose of ulcers, at which point he confessed to his wife what he had done (well, I assume he went home first and didn't just turn to her on the hospital bed with a tube hanging out of his arse and say 'Listen honey, there's something I need to tell you').

After two weeks of suffering, Marshall's wife insisted[37] that he seek a full course of treatment to get rid of the ulcers – his breath was apparently *that* bad – but Marshall had finally proven ulcers were the result of bacteria and could be treated as such.

Doctors all over the globe decided to give antibiotics a go and discovered they were successful. In fact, serious ulceration is no longer a life-threatening illness.[38] Marshall is almost single-handedly responsible for the decline in global stomach cancer rates and has saved countless lives. In 2005 he was awarded a well-deserved Nobel Prize. And now, onto a slightly more pleasant beverage than human stomach juice . . .

Bag it up

In 1908 the New York tea merchant Thomas Sullivan began shipping tea leaves around the world in silk pouches because it was cheaper than using tin boxes.

He intended for his customers to open the bags and pour the tea leaves into their strainers, but apparently they didn't realise that was the point and thought the silk pouches were the delivery mechanism, so they simply began dunking them in hot water.

Since the molecules in tea are small, they can diffuse out of a silk bag without a problem – and thus was born the teabag. There was already a patent for a similar design registered in 1901 to Roberta Lawson and Mary Molaren, but Sullivan's tea empire was already established and he ended up getting the credit for the teabag's accidental invention.[39]

A CORNY STORY

The most famous of all accidental foods was invented by the American broom salesman John Kellogg (you probably know where this is going). One night in 1894 John was called to help his brother William at the local psychiatric hospital.

He had been preparing some wheat dough and left it on the side to prove while he was out. For those of you who've never seen *Bake Off* and witnessed the delightful ogre-ish charms of Paul Hollywood, 'proving' is a stage in bread-making when you let the dough sit, giving the yeast a chance to start fermenting the sugars, thus proving they are alive.

John Kellogg left his dough overnight while helping his brother, and the dough massively over-proved in his absence. The yeast broke down all the sugar to carbon dioxide which escaped, leaving the dough to collapse. Rather than chucking it out the next morning, the brothers Kellogg decided to pass it through a roller and found it broke into little flakes which could be cooked.

At some point one of the brothers (or potentially John's wife Ella) had the inexplicable idea to try eating these flakes in bowls of milk . . . because that's what you'd do, obviously. Eventually, the Kellogg brothers discovered that corn wheat gave the most satisfactory flakes, and a cereal empire was born.[40]

Hopefully not for breakfast

In 1530 a group of monks from the Saint-Hilaire monastery in southern France were making bottles of wine. The process of fermentation was well understood by this point: add yeast to sugar (usually obtained from fruit juice) and leave it for many months. The yeast consumes the sugar and excretes both ethanol and carbon dioxide. The carbon dioxide escapes as a gas, while the ethanol is the consumable alcohol.

This is a process we've known for at least thirteen millennia and it was almost certainly discovered by accident because nobody would deliberately stick a bunch of fruit juice in a jar with fungus and hope for the best. But the monks of 1530 ended up making a discovery.

The winter that year was especially harsh and from what they could tell, the cold weather had killed off their yeast. This meant they would end up with a much weaker, overly sweet batch of wine with most of the sugar in the bottles left unprocessed. They bottled it anyway, anticipating a lot of negative feedback on their Yelp page, but when the summer of 1531 came around the bottles started exploding. Something very peculiar was going on.

What the monks hadn't realised was that the yeast had not been killed by the winter cold but had become dormant in hibernation. It was still present in the wine and very much alive. When summer began warming the cellars, the yeast reactivated and began consuming the sugar it was floating in, producing alcohol and carbon dioxide in the process. But this time the liquid was already sealed inside a bottle, leading to a build-up in gas pressure which ultimately popped open the glass containers.[41]

The monks considered this batch a disaster, but it turned out people actually enjoyed the fizzing taste. By making bottles from thicker glass, they were eventually able to force the gas to remain dissolved in liquid, only to foam out when the bottle is opened. It was the monk Dom Pierre Perignon who began mass-producing the sparkling wine from his own recipe in the Champagne region of France, giving the celebratory drink its name.

PART TWO

Misfortunes and Failures

'He picked up the lemons that Fate had sent him and started a lemonade stand.'

Elbert Hubbard

'That which does not kill me, can only make me stronger.'

Friedrich Nietzsche

'Life's not fair'

Scar (*The Lion King*)

THE HEADACHE TO END ALL HEADACHES

Neuroscience emerged during the nineteenth century, pioneered by the German physicist Gustav Fechner, who only got interested in the subject after he went partially blind from staring at the sun for several months to see what would happen.[1]

It was while trying to make a recovery in bed that Fechner decided to explore the link between the brain and the mind, believing that one directly affected the other. His ideas became what are today the fields of psychiatry, psychology and neuroscience,[2] and the first significant – albeit gruesome – proof that he was right came from an unfortunate railroad accident.

In 1848, Phineas Gage was laying explosives to clear ground for a new railway track through Vermont, a procedure which is achieved in the following way: first a hole is bored into the rock and filled with blasting powder. Second, the powder is topped off with sand before a long pole called a tamping rod is shoved down to pack the explosives as deeply as possible.

On 13 September 1848 at around half past four, Gage was tamping the blasting powder when something caused it to ignite. Nobody knows what caused the spark but the explosive went off too soon and shot the tamping rod straight through Gage's head.

It entered below his left eye, emerged through the top of his skull and sliced a cylinder of brain out in the process. The rod had a diameter of 3 centimetres and thus took out a corresponding piece of Gage's left frontal cortex.

Miraculously, Gage did not die – most likely because the metal was so hot it cauterised the wound and stopped him from bleeding out. As his co-workers watched in horror, Gage stood up, complained of a headache (no kidding) and saddled his horse to ride into town. When he got there, forty minutes later and still alive, he calmly introduced himself to the physician Edward Williams and said: 'Doctor, here is business enough for you.'[3]

The next month was difficult for Gage as he kept slipping in and out of consciousness and his doctors had to repeatedly sterilise the wounds in his skull, but gradually he returned to something resembling health.

Unfortunately the biggest question of all surrounding Gage – 'How was this possible?' – is unanswerable. Medical records in the mid-nineteenth century weren't brilliant, and even if we had flawless accounts of everything, we don't know enough about the brain to answer it.

Somehow, Gage continued to live after having part of his brain destroyed, but in the aftermath, his doctors and co-workers noticed something disturbing. Gage didn't seem like himself any more. Again, this might fall into the 'no kidding' category, but it was deeper than that. Gage's personality had changed altogether.

He wasn't made to take IQ tests, nobody gave him Proust's questionnaire and he didn't complete one of those 'Which cupcake are you?' quizzes, so we only have a handful of records to go on – but what little these records tell, tells a lot.

According to John Harlow, the principal doctor who tended Gage, before the accident he was hard-working, capable and well-liked by other members of the railroad team. After the accident he was 'gross, profane, coarse and vulgar . . .' – he even began swearing and became lazier, abandoning tasks if they were too demanding.[4]

This seemed to point to something inescapable – the brain was the seat of personality, and who you were could be altered by changing it. Before Gage, it was widely assumed that human thoughts and personalities were separate and transcendent to the body, but Gage's dramatic behavioural shifts made this hard to justify. If there really was an immutable soul, how come it changed when your brain got compromised? Shouldn't it be unaffected by something as base and material?

Gage's case is sometimes exaggerated, making out that he became a philandering psychopath. While this isn't true, the fact remains that the part of the brain he lost was the part which had made him a likeable man.

The most common hypothesis is that the left frontal cortex of the brain is the region associated with impulse control and delayed gratification. By removing that, Gage was reduced to his cruder instincts. We all have a selfish part of ourselves that tells us to be rude and do things half-arsed. The frontal cortex is the bit which intervenes and makes us do better. By losing that, Gage lost the ability to 'do the right thing' and turned into a jerk. At least . . . for a while.

One of the most fascinating parts of Gage's story, a part that is often missed, is that in later years he returned pretty much to normal. By the late 1850s he even became a respected and well-liked stagecoach driver whose only problem was that he was blind in his left eye.[5]

This is one of the first records of 'neural plasticity': the brain's ability to change, repurpose and regrow its functions. Gage may have lost the ability to be a good person after the accident, but somehow his brain knew it was an important skill to have and re-trained itself.

Freeze frame

On 6 August 1860 a stagecoach being driven through the Texas Cross Timbers prairie failed to brake and crashed. Among the passengers was a mild-mannered English book salesman named Edward James Muggeridge, who was thrown from the coach and landed headfirst on a rock.

When he regained consciousness, Muggeridge found he was seeing two of everything because his brain was no longer able to merge information from his left and right eyes, giving him a condition called cross-diplopia. In addition, he had lost his sense of taste and smell, despite no injuries to his nose or mouth.

He returned to England to make his recovery and was treated by William Gull, Queen Victoria's personal physician (who some suspect of having been Jack the Ripper), but once his pain had subsided, the residual changes in his behaviour were notably strange.

For one thing, Muggeridge insisted on changing his surname to Muygridge to Muybridge to Maybridge, his middle name from James to Santiago, and his first name from Edward to Eduardo to Eadweard. He also became increasingly aggressive and shot his wife's lover in an act of premeditated murder, but was acquitted on the grounds of justifiable homicide facilitated by brain damage – one of the earliest examples of an insanity defence.

More importantly though, the accident gave him an obsession, which consumed him for the rest of his life. During the crash, Muybridge noticed that his experience of time changed significantly. He perceived events moving at a slower rate, so slow, in fact, that everything seemed eventually to stop and become still. He could recall the accident so clearly it was as if he was moving from still image to still image.

The concept of things moving in 'slow motion' seems mundane to us because we live in a world of YouTube playback speed options and Zac Snyder movies, but in 1860 the idea of time slowing sounded like madness. Nevertheless, the accident had given Muybridge an idea: if you slow moving objects down, everything becomes still, so would it be possible to reverse the process? Could you take a series of still photographs and play them back fast enough to create movement?

Muybridge set about building a primitive version of what eventually became the motion picture camera, although the design he used would barely be recognisable today and he only got the funding to do it thanks to a bet. Here's what happened . . .

The University founder Leland Stanford wanted to settle an argument. When a horse gallops, is there a point at which all four hooves are raised off the ground? Muybridge claimed he could find the answer. With Stanford's help he rigged up a row of cameras along a stretch of land on the latter's ranch in Sacramento, each with a trip wire that would capture an image of the horse a moment after its leg struck the wire.

Muybridge was able to capture a sequence of photographs showing that there was indeed a moment when all four of the horse's hooves were off the ground. His photographs could then be projected onto a screen and rotated through at high speed. Thanks to the brain's inability to move from still image to still image quickly (an optical phenomenon known as 'persistence of vision') the illusion of movement was created.

In fact, one of the earliest pieces of film footage you've probably come across (an unnamed jockey riding a horse) was recorded by Muybridge in an attempt to develop his technique.[6] Would we have movies if Muybridge hadn't been in that accident? Nope.

LIGHT THE CORNERS OF MY MIND

An even more unusual psychiatric misfortune happened a century later in 1953 when an American named Henry Molaison underwent a procedure to rid of him of epileptic seizures he'd had since a childhood bicycle accident.

The doctor performing the surgery had learned that removing parts of the hippocampus in psychotic patients calmed their seizures, so he tried it on Molaison as a last-ditch attempt. When Molaison recovered from the surgery he had indeed stopped having seizures, but what nobody expected was that he also stopped forming new memories.

For the rest of his life, Molaison was convinced he was still a twenty-seven-year-old man living in the year 1953, even when he was well into his eighties. He was insistent he was in the prime of youth for the rest of his life and yet was incapable of learning anything new from the day of the surgery onward.[7]

What had happened to him? Very slowly his doctors started to realise that the hippocampus is the brain's central processing unit for memory. While the electrical impulses that create a memory are stored all over the brain, the hippocampus acts as a coordinator. Damage that and you end up with a patient incapable of learning, like Guy Pearce's character in the film *Memento*.

We've still got a long way to go before we can fully explain how a bag of salty water and fat inside our skull is capable of storing data, but the Molaison surgery was the first clue that certain regions of the brain are in charge of storing sensations. The very notion of learning information would eventually be explained because we accidentally removed one man's ability to do it.

A similar case is recorded of a man codenamed 'NA' who also received an injury to the hippocampus which destroyed his ability to form new memories. We know the injury took place sometime during 1960 in Portugal when he was twenty-two, but little else – apart from one of the most cryptic statements in medicine. We are told only that 'he sustained a penetrating brain injury with a miniature fencing foil'.[8]

DRASTIC GASTRIC

In 1822 a Canadian fur trader named Alexis St-Martin was accidentally shot in the torso with a musket. St-Martin was assumed to be a goner, but thanks to the skilful surgery of a doctor named William Beaumont who rode on horseback to his rescue, St-Martin made a recovery. With one curious side-effect – he developed a tube connecting his stomach to the surface of his skin.

By a freak accident, St-Martin's cells had ended up reconstructing his anatomy with a mistake and built a tunnel from his stomach to just under his left nipple. His body had a second, unintentional oesophagus.

William Beaumont immediately spied an opportunity and, rather cruelly, got the illiterate St-Martin to sign a contract giving Beaumont permission to carry out anatomical experiments on him. Almost nothing was known about digestion at the time and this was a chance to watch it happening.

Beaumont began by tying food on bits of string, sliding them down St-Martin's chest opening and pulling them back out at intervals to see how much had broken down. Sometimes he would try to peer into the hole with a lamp and watch what was going on; sometimes he would put food in little sacks to see how effective the acids and enzymes were at getting through the cloth.⁹

Over the course of ten years (during which St-Martin tried to escape several times but was always brought back by the over-zealous and uncaring Beaumont) over two hundred experiments were performed. Unfortunately a lot of them proved useful.

Beaumont was the first person to discover that the stomach contained hydrochloric acid, that the stomach pulverises food, that different foods break down at different rates depending on ingredients and – most surprisingly – that the mood of a person will affect their rate of digestion: St-Martin would digest his meals slower whenever he was angry (which he often was).[10]

WHO EVEN *WAS* DR TUGGLE?

Just to balance things out with a more upbeat story of torso injuries lead-ing to discovery, consider the life of John Pemberton.

Pemberton was stabbed through the gut with a sword during the American Civil War and became addicted to morphine to dull the constant pain. Not happy with the inevitable side-effects of morphine addiction, Pemberton decided a better way of administering painkillers had to be found and set about inventing one which didn't contain opioids.

His first attempt was a tonic he called 'Dr Tuggle's Compound Syrup' which did stop the pain, but its active ingredient came from Alaskan buttonbush flowers which are toxic. Not a great alternative.

Undeterred by the set-back of making a poisonous painkiller, Pemberton invented a new recipe which blended cocaine, wine, kola nuts and damiana flowers. Unfortunately, he was thwarted once again when temperance laws hit the USA and wine-based drinks became ille-gal (cocaine was fine though). His new alcohol-free drink wasn't anywhere near as good, especially when warm, so he needed a way of cooling the drink down.

He went into business with carbonated-water magnate Willis Venable and together they devised a solution. When you bubble carbon dioxide through water the added gases prevent the water molecules clumping together which means they can't hold heat so well. Carbonated water is therefore easier to cool than regular water. You just have to put up with the fizzing side-effect.

Venable and Pemberton started mixing carbonated water with the cocaine and kola-nut juice, only to discover that people loved the fizzy taste. With a little marketing, Pemberton's potion became Cocaine-Kola and eventually Coca-Cola™, the most popular soft drink in the world. All it took was prohibition laws, poisonous plants, morphine addiction – and getting stabbed briefly in the stomach.[11]

Hard science

In the early 1990s, chemical supergiant Pfizer were looking to test a new drug they'd developed for angina – chest pain caused by poor blood flow to the heart. It had been devised by lab workers in the picturesque English village of Sandwich and computer models had shown it might help bind to key biological receptors that malfunction in angina sufferers. There were promising results in animal trials so Pfizer turned to one of their leading researchers, Ian Osterloh, to conduct a human study at his hospital in Swansea, South Wales.[12]

Codenamed UK-92480 the medicine was trialled on a group of angina sufferers, but after several months it proved a complete failure. While it did relieve *some* symptoms, it wasn't anywhere near strong enough to be clinically significant or even much better than a placebo. After all the money and time Pfizer had spent, UK-92480 turned out to be a waste. This is very common in medical research – but what happened next is really quite rare.

Things became curious when one of the nurses monitoring the group of test subjects reported unusual behaviour in the male patients. After being given the drug she found each of the men lying on their stomachs, refusing to be interviewed standing or sitting. A bit of gentle questioning revealed the reason. They were embarrassed because *all* of them had suddenly found themselves with unquenchable erections and they didn't want to be rude.[13]

Initially, Osterloh wasn't interested in this side-effect, but the more he administered the drug the more consistently he recorded it. He and his team had discovered a treatment for erectile dysfunction, which is a big deal both medically and financially. While 15 per cent of men suffer from angina, erectile dysfunction moves in lockstep with age group, i.e.

40 per cent of men experience it in their forties, 50 per cent in their fifties, and so on. UK-92480 was repackaged as an erectile dysfunction medicine and you probably know it by its trade name: Viagra™.

The full science of erections is complicated but the basics are straight-forward. First, something happens in the brain to trigger sexual excitement. Second, the brain causes the release of a chemical called cyclic Guanine MonoPhosphate (cGMP for short) which is in charge of getting muscles to relax. You might think relaxation is the opposite of what you want here, but certain types of tissue allow blood to flow through them when relaxed, causing expansion.

But alas: we can't just swallow cGMP and hope for the best. cGMP has a lot of additional functions involving retina health and memory formation. The body has thousands of systems which direct chemicals where they're needed, so simply upping the body's cGMP would affect those pathways too. Instead of ingesting cGMP, we need to make sure the body's existing cGMP pathways are working better.

cGMP has an arch nemesis, PhosphoDiEsterase-5 (PDE-5), which has the job of breaking it down when it's building up. The problem with erectile dysfunction seems to be that the body is over-producing PDE-5, which destroys the cGMP before it has a chance to do its job. This is what a blood dilating medicine needs to work on. Rather than attacking the symptom, i.e. 'not enough cGMP', a successful medication will treat the cause, i.e. 'PDE-5 breaking it down too much'.

In simulations, compound UK-92480 had the ability to stick to PDE-5 and stop its onslaught against the much-needed cGMP. In theory, they thought, UK-92480 should allow blood flow in the heart to return to normal. And they were correct – just not in the place they expected.

Precisely *why* the compound works best in the penis rather than the heart is not known (insert your own joke about the way to a man's heart here). There may simply be more PDE-5 in the penis because penises need to break down cGMP more often – you don't want to have an erection constantly, after all. You have to go outdoors sometime.

The first trial of Viagra™ for treating erectile dysfunction involved a dozen men, and it helped them all achieve erections as doctors showed them pornography (in the name of science, you understand).[14] Then, a trial of over two hundred men found that in 69 per cent of them (I didn't make that number up) Viagra™ was effective.[15]

This was a huge deal because prior to Viagra™ there was only one way to artificially stimulate an erection, and it was a bit grim. Professor Giles Brindley of Cambridge University had discovered in 1983 that two chemicals called papaverine and phentolamine were able to dilate blood vessels, but they had to be injected directly into the penis to work. Something he once demonstrated in eye-widening fashion.

At the 1983 Urodynamics Society conference in Las Vegas, Brindley was due to give a lecture and decided he would use it to unveil his discovery. In front of a hall of distinguished urologists, Brindley shuffled onstage in a tracksuit and treated his audience to photographs of his own erect penis. Thirty photographs.

But this wasn't enough.

After giving his presentation to an increasingly silent crowd, Brindley explained that the slideshow didn't prove those erections were medically induced. So, to demonstrate the effectiveness of his treatment, he pointed out that giving a lecture was very un-sexual, yet he was sporting an erection at that very moment because he had injected his penis with papaverine in his hotel room.

Brindley then stepped out from behind the podium and revealed why he was wearing a tracksuit. Tightening the fabric around his crotch he demonstrated that he was erect.

But this wasn't enough.

Brindley decided the tracksuit pants were muting the effect and his erection wasn't properly visible (one suspects the audience was willing to take his word for it at this point). He decided the only way to prove the erection would be to pull the tracksuit pants down in front of everyone. Which he did.

But this still wasn't enough.

After basking in a pin-drop (and pants-drop) silence Brindley decided that to *really* prove his point he would ask the audience to inspect his penis for stiffness. So he started waddling down the stairs of the podium, brandishing his erection in front of him like a missile carrier. Screams erupted from the front row as doctors scrambled away in horror and it was at this point he decided he had gone far enough and returned to the stage. Naturally, little else from this conference is remembered.[16]

Brindley published his results six months later in the *British Medical Journal*, but there weren't many takers for the treatment. Viagra™ provided a more favourable solution for men who didn't fancy sticking a syringe needle into their penis – which is *all men*.

An important qualifier is that Viagra™ only boosts a penis's erection if there's cGMP present to begin with, i.e. if you haven't got any, inhibiting the PDE-5 will be useless. Viagra can't give a man an erection out of nowhere and it doesn't increase libido or stamina.

There has to be some sexual stimulus to get things going so taking Viagra™ won't give you an erection if all you do afterwards is look at gardening equipment (well, it depends how sexy the gardening equipment is). What it *will* do is facilitate your already existing sex drive. Viagra™ today is valued at an estimated $2.5 billion.

THE SHAPE OF THINGS TO COME

As a final word on unintentional penis science, in 1999 the Dutch researcher Pek Van Andel was wondering what was going on inside the body during an orgasm and decided to get couples to have sex in an MRI scanner while he watched.

Thirteen experiments were conducted on volunteer couples, who had to hold still mid-orgasm for the machine to get a clear image. Aside from some unsurprising results (it was difficult to perform in an MRI scanner) there was one very surprising one. It turned out that during intercourse an erect penis takes the shape of a boomerang, with at least two thirds of it curving inside the body.

In a way this was exciting news because it meant all men could suddenly claim to have a much bigger appendage than previously thought, but it was also quite humbling to realise that, for most of human history, we'd been getting the shapes of penises wrong.[7]

ROYAL GUNK

In 1856 the German chemist August von Hofmann was trying to make quinine, an effective treatment for malaria. According to Peruvian legend, the discovery that quinine could be used this way was made by a young man suffering from crippling diarrhoea in one of the Andean jungles. The dehydration was threatening to kill him, so in desperation he drank from a pool at the base of a cinchona tree. Cinchona was thought to be poisonous, but the man had no choice. What he didn't realise was that the bark of the cinchona tree contains quinine, which had seeped into the groundwater. Within days his fever faded and he had found a cure.[18]

Hofmann was hoping that if his lab could find a way to synthesise quinine artificially it would be possible to mass-produce it, rather than having to grow and harvest cinchona trees. If the chemical could be manufactured it would save hundreds of thousands of lives.

In the mid-nineteenth century there was no real theory for chemical reactions, however, and no understanding of why one substance will turn into another. We hadn't discovered atoms, only 50 per cent of elements had been isolated and the Russian chemist Dmitri Mendeleev wouldn't invent the periodic table for another thirteen years (see The periodic table of 'huh?' at the end of the book).

It would also be another seventy years before Linus Pauling discovered the nature of the chemical bond – a discovery which is to chemistry what evolution is to biology (and the fact Pauling's name isn't on a plinth in every chemistry lab in the world is a freaking outrage). But I digress . . .

The point is that in the nineteenth century, reactions were mostly a combination of guesswork and bucket chemistry, i.e. chuck a bunch of substances together and hope for the best. Hofmann was mixing chemicals with similar properties to quinine and hoping one of them would be close enough to turn into quinine proper.

When the Easter holidays rolled around, Hofmann decided he needed a break so he handed the research over to his eighteen-year-old lab assistant William Perkin as a homework project, presumably saying something along the lines of: 'Come along Perkin, be a good lad and discover the cure for malaria won't you?'

Perkin had a small laboratory in his attic and decided to take all the necessary chemicals home with him, cursing Hofmann's name under his breath as he did so. Over the Easter break, he set about mixing the various potions in every order he could think of, hoping one of them would have the right density and appearance, but had no luck[19] until – for no real reason – he reached for some aniline and mixed it with potassium dichromate. Aniline is an oily liquid which smells like rotten fish and can be extracted from boiling leaves of the Indigofera plant. Potassium dichromate is a bright orange liquid with no odour. Both are highly toxic.

Unfortunately for Perkin the aniline he was using was not high quality and contained an impurity called toluidine. Toluidine is often found as a contaminant in samples of aniline, but this sample must have been truly horrendous because the aniline was close to *half* what was inside the bottle. If you have a jar of toluidine and aniline, the two chemicals will react together, especially in the presence of potassium dichromate. The result was a thick black sludge glued to the bottom of the reaction flask. Failure again.

This Marmite™-esque blob is a common sight in labs and it means that your starting molecules have uncontrollably and randomly reacted, sticking together all willy-nilly. This material is good for nothing (except showing to children who are thinking of smoking), so Perkin sighed and moved to the wash-up station.

The standard method of cleaning out black tar from glassware is to throw the glassware away and apologise to the lab technician, but Perkin decided he was going to salvage the flask. He was on holiday after all, and had to make do with what he had. So he poured in a solution of alcohol to dissolve the tar . . . and suddenly it turned a rich and beautiful purple.[20]

This was exciting. The colour purple was famously the most difficult shade to obtain in both the art and textiles industries. Red dye could be obtained from cochineal beetles, yellow dye from goethite rocks and blue dye from spirulina fungi, but purple dye was elusive. One of the only sources was a species of Italian sea snail called *Murex brandaris* and the purple it produces is pale, meaning you need thousands of them to dye a single piece of fabric.[21]

Subsequently, purple dyes had, since the time of the ancient Greeks, been highly sought after. Only the wealthiest could afford to have purple fabric and owning some was considered a symbol of great status and influence. Even today, the British royal family use purple for their formal robes because purple is the most regal of colours.

Nobody had a method for generating purple colouring agents – except for this one eighteen-year-old standing in his attic, staring into his reaction flask, black slime coating his hands. Perkin had discovered a technique for making the most valuable colouring agent in the world.

The chemical he had made, which he originally called 'Tyrian purple' but eventually gave the catchier name 'mauveine', was the result of reacting aniline and toluidine. It's quite probable that other chemists had manufactured mauveine themselves and simply not realised because they saw the black mess and chucked it away. But because Perkin decided to clean his glassware he ended up dissolving the tar and revealing the dye inside.

With the help of his brother, Perkin set up a secret lab in his garden shed where Hofmann wouldn't come looking, and together they figured out a way of manufacturing mauveine in a controlled manner.

Hofmann was not happy when he found out what was going on and described Perkin's discovery as 'purple sludge'. Perkin, however, sold the patent and became a millionaire. Even Charles Dickens wrote an editorial praising the dye for the way it highlighted women's features: 'O Perkin's purple, thou art a lucky and a favoured colour.'[22]

Although nobody knew *why* mauveine formed, they knew that using similar starting chemicals might yield other colours. And they did. Abundantly. By the end of the nineteenth century, Perkin's technique had generated over ten thousand artificial dyes.[23]

The understanding came much later that the cause of colour in a chemical is down to a molecular feature called a pi-bond system. One of the most common pi-bond systems in nature is a benzene ring – a hexagonal loop of carbon atoms* – and both aniline and toluidine contain them. Mauveine is simply the result of these molecules sticking together.

This was the key feature needed for a synthetic dye to work. By using benzene-based molecules, people increased the chance of making something which had colour. It was still bucket chemistry, but Perkin had identified that one of the buckets needed to contain benzene-rich molecules and today the synthetic dye industry is worth $6 billion.

---

* Benzene's structure came to the German scientist August Kekulé while he was dozing in a horse-drawn cart and had a dream about a serpent eating its own tail, giving him the idea for a loop-shaped molecule.

STICKY BUSINESS

There are three main ways objects can stick together: mechanical, electromagnetic and chemical. Mechanical is the simplest and involves two objects becoming entwined so that friction prevents them from sliding apart. This forms the basis of things like knots and hook-loop fastening fabrics, which were invented in 1941 by Swiss electrical engineer George de Mestral when he noticed burdock seeds sticking to his dog's fur[24] (the rules for even writing the name of this product are so complicated I haven't been able to figure them out, so to avoid getting sued I've just called it hook-loop fastening fabric – you know, like the stuff little kids use to do up their shoes before they've learned laces).

The two other forms of bonding are more complicated, and the most famous application of both are the result of accidental discoveries.

In 1942 Harry Coover was working in the research and development department for Eastman Kodak, New York, when he was approached by the military. They wanted him to manufacture a transparent plastic which could be used for gun sights. Most gun sights are made of metal in order to handle the heat of a bullet explosion, but a clear sight would obviously be preferable.

Coover and his team set to work researching a group of molecules called esters. Esters are easy to bond to each other so it's possible to create chains of them which can be tangled together to form plastics. While studying various esters, Coover's team ended up making a new one called methyl-2-cyanoacrylate. Unfortunately, however, rather than forming a rigid plastic, methyl-2-cyanoacrylate was a ludicrously sticky goo which bonded to everything it touched, so the Government abandoned the project.

Six years later, Coover was trying to develop another kind of plastic for an aeroplane canopy when he decided to revisit his super-sticky concoction. He gave it to one of his lab assistants, warning him that any equipment he tried testing it with would become stuck. A few weeks later his assistant told him he'd been having fun randomly sticking everything he could think of to everything else. Coover realised at this point that what he had initially considered a failed plastic was actually the world's most powerful glue.

After showing it to his superiors, Kodak immediately saw the application and began selling it as 'Eastman 910 adhesive', later rebranded as Krazy Glue™ and Super Glue™. Coover even demonstrated the strength of his formulation on the TV show *I've Got a Secret* when he lifted presenter Gary Moore on a cable, using a single drop of glue.

During the Vietnam War, a US general approached Coover and asked him about using Super Glue™ as a surgical tool for patching up wounded soldiers on the battlefield. While it was not officially sanctioned (and the name of the general is not disclosed) Coover discovered that by spraying Super Glue™ onto an open wound it could seal flesh quicker than stitches, leading to countless lives being saved.[25]

The key to Super Glue™ is that cyanoacrylate molecules are not sticky on their own. They won't bind to each other or the inside of the tube but as soon as you bring them into contact with water, they begin coagulating. Think of the way dry flour works: it's not terribly sticky, but turns into a paste when wet.

Water molecules, found on the surface of just about everything, trigger cyanoacrylate molecules to self-bond, forming chains which are hard to break. If the cyanoacrylate glue is sitting between two surfaces as it starts clumping together, it will form bonds to both surfaces, fusing them in the process.

Naturally, Super Glue™ works best on materials that have high water content (like human skin), but even dry materials can be bonded with moisture from the air. The only way to get the glue off is to dissolve the chains with solvents. Adding water isn't going to help.

SLIGHTLY LESS STICKY BUSINESS

What if you don't want something to be permanently sticky? What if you want something to be tacky enough to hold things together, but not so much that it causes permanent adhesion? For that, we turn to the American engineer Spencer Silver who, in 1968, was working for 3M trying to make a new type of glue for aeroplane metal.

Super Glue™ was great, but not ideal for metal surfaces which usually have low water content. Silver was testing different formulations for glues, and instead of making one that was extremely strong he ended up making one that was extremely weak. His recipe contained acrylate molecules – the same as Coover's – but the resulting compound was clearly useless. Who wants a glue that doesn't glue things?

Silver tried convincing his bosses that a mild adhesive could still have applications, e.g. he proposed making blackboards with a coating of the glue, allowing you to stick paper on to it, but nobody was interested. For years, his invention circulated at 3M as a curiosity or, as he described it, 'a solution without a problem'.[26]

Eventually he showed his weak glue to an engineer named Arthur Fry who immediately thought of a use. Fry sang in his church choir but found the paper bookmarks he used would slide out of his hymn book (hymnal paper is very glossy). He decided to attach Silver's adhesive to a piece of paper and use it as a slightly sticky bookmark. It worked perfectly.

Silver had been pitching the idea backwards all these years. Rather than having a sticky surface you press paper onto, what if you had pieces of paper with a thin strip of sticky stuff on them? A little bit of fiddling to make sure the adhesive stayed on the paper and 3M floated the product as the Press 'n' Peel™, before eventually marketing it more widely as the Post-it™.

The mechanism of Post-it™ notes is more complicated than Super Glue™ because, rather than sticking to something by chemical reaction, they work through electromagnetic attraction.

Every atom has electrons whizzing around the outside. As they move from one side to the other, they turn the atom into a tiny magnet with an electron-rich end and an electron-lacking end. The atom next door will then flip to generate the corresponding magnetic field in response, bringing the atoms closer. This process never stops, which means if you bring any two atoms together there will be a faint attraction between them. We call these attractions 'London forces'.

Most of the time we don't notice London forces because other forces overcome them by a factor of millions. If you were to put your hands on the underside of an aeroplane you wouldn't get carried away as it took off, even though technically your hand is attracted to the plane. Gravity and air resistance are always going to win that fight.

You can just about feel the effect of London forces if you press your palm firmly against a smooth table. When you try to lift your hand, you can sometimes feel a weak 'tug' as the table holds onto you. What's happening is that your soft skin has spread out and its surface area creates just enough points of contact between your atoms and the table's atoms to create attraction.

Geckos use this better by having wrinkled hands with an enormous surface area, leading to greater London forces. Spiders use a similar technique to walk up walls. They are so light that the tiny electromagnetic attractions between them and the wall are strong enough to overcome their weight. Post-it™ notes work the same way.

The material on the back of a Post-it™ is made of long strands of acrylates which bunch up into balls. When the paper is pressed onto something the bunched-up strands spread out and fill the holes of whatever surface you're applying it to. When you stop pressing, the strands have coated the surface and, by filling the holes, have increased the London forces. Not strong enough to hold things together permanently but strong enough to make them cling.

Post-it™ notes won't stick to everything of course, especially if the surface is smooth, since there's not enough surface area for the strands to touch, but anything rough will work. Today Super Glue™, a failed plastic, is worth $2.5 billion while Post-it™, a failed glue, is worth $2.3 billion.

PLASTIC FANTASTIC

Polythene is one of the simplest plastics, consisting of carbon atoms strung together in a chain with two hydrogen atoms bonded to each one like so . . .

$$
\begin{array}{cccccc}
H & H & H & H & H & H \\
| & | & | & | & | & | \\
C - & C - & C - & C - & C - & C \\
| & | & | & | & | & | \\
H & H & H & H & H & H
\end{array}
$$

It's now known to cause all sorts of environmental problems because it's so strong it doesn't easily degrade and fibres of it can stay in the ecosystem for hundreds of years. We're therefore trying to move away from using polythene, but for decades it was the most beloved plastic in the world (perhaps second only to silicone). And it was discovered by accident. Three times.

In 1894 the German chemist Hans von Pechmann discovered a new chemical named diazomethane, a fatally toxic yellow gas used as an intermediate for making ethers. But when Pechmann dissolved it in solution he noticed annoying little white flakes appearing.

While writing to his chemist friend E. Hindermann (whose first name is lost to history) he told him about the flakes and Hindermann wrote back with astonishment – he had found the same thing!

Diazomethane formed annoying white precipitates when you dissolved it.

Both chemists decided to disregard these flakes as a by-product because they were interested in diazomethane, not realising they had both independently created and discarded what would become the most widely used plastic in history.[27]

The third accidental discovery occurred in 1933 when Reginald Gibson and Eric Fawcett were working with a simpler chemical called ethene at Imperial Chemical Industries in Cheshire. Ethene consists of two carbon atoms bonded to each other, with two hydrogen atoms protruding from each one, forming a crude letter 'H' shape.

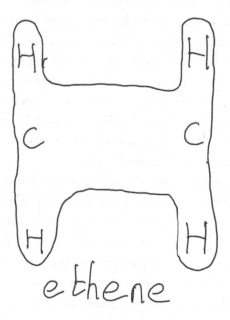

ethene

Gibson and Fawcett were trying to react ethene with benzaldehyde in the hope of manufacturing a new industrial resin. On 24 March they loaded the two starting chemicals into the pressuriser, left it to churn over the weekend and went home. But there was a problem. Their apparatus had a leaking valve. A tiny bubble of air got into the reaction mixture and destabilised the ethene molecules, causing them to bond with each other rather than the benzaldehyde, forming chains of poly-ethene (polythene).*

ethene ethene

Poly-ethene

When Gibson and Fawcett returned to the lab on Monday they found the reaction chamber had become clogged with a white substance that would not budge. Cursing the faulty valve, they tried to clean the mess out of the machine but found it nearly impossible. The newly formed material was so stretchy and strong that it couldn't be scraped away.

Fawcett reported their findings to a conference on polymer chemistry later that year but nobody believed his claims, including the eminent Nobel-Prize-winning chemist and polymer expert Hermann Staudinger. There was just no way Fawcett and Gibson had polymerised ethene – and even if they had, what use would it be?

It was another two years before Fawcett and Gibson's claims started to be taken seriously and by the Second World War, their plastic was being used in the manufacture of radar equipment. Radar sets up until then were bulky and heavy, making them impractical to install in fighter planes. But with the invention of polythene as a lightweight insulator, the British were able to outfit the Air Force with radar devices capable of locating Nazi U-boats.

This gave the Allied Forces such a potent advantage that Hitler's second-in-command and brief successor Karl Dönitz lamented that British radar was so good it risked Nazi defeat.[28] Perhaps it was fortunate that Hermann Staudinger hadn't taken the polythene claims seriously after all.

* Oxygen molecules, when heated under pressure, split into unstable atoms called free radicals. These free radicals attack the double bond of an ethene molecule and cause it to pop open, leading to reactions between the ethenes.

SLIPPERY WHEN DRY

Something very similar happens if you switch hydrogen atoms for fluorines, as was discovered by accident in 1938 by American chemist Roy Plunkett. Plunkett was working with tetra-fluoro-ethene (TFE), a molecule made up of two carbon atoms with four fluorine atoms in the positions we previously saw for hydrogen.

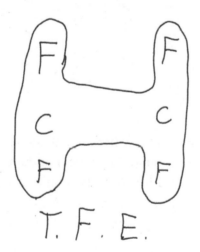

T. F. E.

Plunkett was collecting TFE gas in an iron cylinder which he then put under pressure. He was actually carrying out an unrelated experiment which only needed TFE to carry heat away in short blasts. The simplest way to make jets of TFE is to put it under pressure and release it through a spout, a bit like puncturing a hole in a balloon. The TFE was only supposed to be a coolant, but when he and his assistant Jack Rebok attached the TFE pump to their experiment nothing came out.

At first they assumed the TFE in the pump had run out, but when Plunkett looked at the scales he saw this wasn't the case. According to the scales, the cylinder was still full, yet none of its contents was emerging. Somehow the TFE was still in there but no longer in gas form.

Eager to solve the puzzle, Plunkett got hold of a hacksaw and cut the TFE cylinder in half, finding the inside coated in a white wax. It turns out that iron reacts with TFE in a similar chain reaction to the one that makes polythene. Plunkett had accidentally made a new polymer called poly-teta-fluoro-ethene – PTFE.[29]

F   F   F   F   F   F
| | | | | |
C — C — C — C — C — C
| | | | | |
F   F   F   F   F   F

This new polymer was similar to polythene except for one key difference. It was notoriously slippery. Plunkett's employer DuPont christened it Teflon™ and it was written off as a joke at first for a very obvious reason: if you want to give something a slippery surface with Teflon™ . . . how do you stick the Teflon™ to the surface in the first place, genius?

The answer came from the French engineer Marc Gregoire, whose wife Colette was getting frustrated by food sticking to her cooking utensils. She asked her husband to find a way of coating frying pans in Teflon™ and he set to work.[30] The solution he eventually arrived at, after much trial and error, was to use the same fragile London forces that give Post-it™ notes their stick.

Since all materials have a weak affinity for one another, Gregoire hit on the idea of blasting the surface of a frying pan with sand to roughen it up. The Teflon™ could then be poured into the micro-cracks, filling them and becoming loosely held in place by the increased surface area.

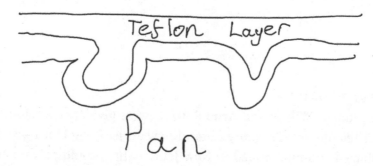

Because the Teflon™ layer is only held to the surface weakly rather than being bonded chemically it's a very loose arrangement, which is why you can't scrub a Teflon™ pan too hard or use metal utensils on it: you'll scrape it right off.

A MESSY SITUATION

In 1964 Robert Wilson and Arno Penzias were given permission to use the Bell Telephone Laboratory's brand-new device – Big Horn – a 50-foot cone-shaped antenna located in New Jersey which could pick up energy pulses from stars. Stars give out plenty of visible light of course, but their low-energy emissions are more useful because they can pass through our atmosphere unhindered, without interacting with any of the atmospheric chemicals.

Wilson and Penzias were particularly interested in the energy coming from stars in the Milky Way but when they began taking readings, something was wrong. They were getting major interference in the form of a weird energy signal at 160.2GHz.

Stars normally give off energy in the region of 2.7GHz so this was not the kind of number they were expecting. Furthermore, its intensity was far lower than what you should get for stars. The stars of the Milky Way are right 'next' to us (cosmically speaking, the Milky Way is still one hundred thousand light years from edge to edge). But these readings were so faint they seemed to be coming from somewhere in the far universal distance.

It would be like listening to music and expecting loud thumping bass, only to get a faint high-pitched whine on the track as well. Something much higher in energy, but fainter in intensity, was interfering with their readings.

At first, Penzias and Wilson wondered if the nearby city of New York was messing with their instrumentation. New York is full of electric lights so perhaps the city's nightly light show was to blame? But that didn't explain how the intensity seemed so far away. It also didn't explain why Penzias and Wilson were getting the readings no matter which way they pointed the dish. Up in the sky, from horizon to horizon and even through the Earth. The obvious conclusion was that something was wrong with Big Horn itself.

When they went to inspect the dish they immediately found the culprit. Two pigeons had set up camp inside it and had liberally deposited a thick white pigeony discharge all over everything.

Penzias and Wilson captured the pigeons and sent them to a bird collector out of state, but unfortunately they were homing pigeons so they flew right back and returned to their pigeon ways. Penzias and Wilson therefore did the only logical thing they could. They shot the pigeons.[31]

However, to their surprise (and hopefully guilt) once they had cleaned all the excrement off the horn, the interference was still there. Frustrated, Penzias began moaning about it on the phone to his friend Robert Dicke at Princeton. When Dicke went silent, Penzias began to twig that there was something important going on.

At that time, there was a huge debate in cosmology over the origins of the Universe. According to one view – the Steady State hypothesis – the Universe had always been roughly the same type of thing. There was no special point in the past when the laws of physics switched on and took their current form. All the basic properties of energy, matter, space and time were steady and eternal.

In the other corner of the debate was the far more radical Big Bang hypothesis which said that billions of years ago the Universe went through a kind of enormous change during which its properties 'began'. Everything was condensed into an inconceivably small point which, for reasons unknown, began expanding.

This idea had been proposed by the Belgian priest and physicist Georges Lemaître, and was endorsed by Pope Pius XII (who misunderstood the idea slightly and thought it was saying the Universe had a clearly defined start, requiring a divine cause),[32] but almost every serious physicist thought it was laughable – including Einstein. 'Big Bang' was a joke-name given to it by one of its chief critics, Fred Hoyle, who said the idea of the Universe popping into existence like a girl jumping out of a party cake with a 'Big Bang' was ludicrous.[33]

To be clear, the Big Bang hypothesis doesn't claim the Universe exploded with a blast. There was no air so it wouldn't have made any sound and it was an extremely small expansion at the start . . . but the 'tiny expansion hypothesis' doesn't sound as cool.

The Big Bang simply states that if you wind backwards in time you'll get to a point where the Universe was so tiny, dense and hot that our physics can't make sense of it.

Robert Dicke was a Big Bang supporter, but he was in a stark minority. He and others had predicted that if the Big Bang *were* true, part of the process of the Universe expanding would lead to a cooling-down period about 380,000 years after the initial stretch. During this cooling down, particles frothing at high energy would start attracting each other and lose energy which would still be detectable today. It would be faint, seemingly coming from a great distance, and would be found in every direction, like a ring in the Universe's ears.[34]

If the Big Bang was true, it should be possible to discover such an energy pulse in all directions, constantly and faintly. Dicke and his team were in the process of making a device which they hoped would make such a reading, but Penzias and Wilson had walked blindly into it, mistaking the origin of the Universe for pigeon poo.

HOLY ETHICALLY TROUBLING STUDY, BATMAN!

In 1959 the Polish-American psychologist Milton Rokeach had an idea. What would happen if you put a group of people who all believed they were Jesus Christ into the same room?

It might sound like a cruel form of human experimentation, but Rokeach's hope was to cure schizophrenics of their delusions. He had read a memoir by Robert Lindner in which two women who each thought they were the Virgin Mary met one other on the lawn of their hospital. Shortly after meeting, one of the women decided there could only be *one* Virgin Mary and concluded that she couldn't be the mother of God after all – apparently becoming cured.[35]

Whether this story of the Marys is true (supposedly it took place in, of all states, Maryland) Rokeach wanted to see if such an approach could cure schizophrenics. He began searching for patients who shared a delusion and at Ypsilanti State Hospital came across three men who all believed they were Jesus.

With the help of two graduate students, Ronald Hoppe and Richard 'Dick' Bonier, Rokeach arranged to have the three Jesuses brought to Ward D-23, where he administered his unconventional therapy in the hopes of breaking their belief cycle.

The identities of the three men were markedly different. Leon Gabor, thirty-eight, came from a family with a history of schizophrenia and he had been raised by a devoutly religious mother. When he decided as a child that he was Jesus and insisted his mother worship him, he was institutionalised.

Joseph Cassell, fifty-eight, had originally been a writer before suffering a schizophrenic break twenty years prior and ending up with a Jesus delusion. Clyde Benson, seventy, had senile dementia and a life of alcoholism behind him, only deciding that he was Jesus in his senior years.

On 1 July 1959, Rokeach introduced the three men, hoping to cure at least two of them but it did not go well. They quickly started arguing over which of them was the real Jesus, sparked by Leon Gabor's claim that his birth certificate proved it.

After initial failure, Rokeach was determined to keep going and instructed his grad students to room the three men next to each other, as well as encouraging them to socialise. Rather than observe the men himself though, Rokeach told his students to document the Jesuses for eleven hours a day. For two years. Classic professor move.

After a few weeks of anger the three Jesuses came to blows, but this willingness to get violent didn't prove anything. The Biblical Jesus became angry on several occasions (Matthew 16:23, Matthew 23, Mark 11:12–14 and John 2:13–22) and it's difficult to predict what his response would have been if he met people who claimed to have his identity. The Biblical records show Jesus believed himself to be the son of God (Luke 2:49, John 8:23) but there are no records of how he responded to someone telling him 'No you aren't, I am'.

The experiment was a tragic misfire because you can't cure people of schizophrenia by exposing them to mutually exclusive delusions in others. Unfortunately, Rokeach was not satisfied with this outcome and began manipulating the situation in a crueller, and more ethically troubling, direction.

He hired an assistant, Mary Lou Anderson, to start flirting with Leon Gabor (the most lucid and articulate of the three). Rokeach wanted to see if Leon could fall in love and be persuaded away from his delusion, i.e. could he consciously choose to abandon a delusion if a woman he loved asked him to?

Sadly for Leon he did start to develop feelings for Mary Lou, as Rokeach hoped, but he was unable to let go of his belief that he was Jesus. This seemed to prove that a delusion cannot be shed even if the person has a clear motivation and desire to do so. Schizophrenics are not in control of their beliefs any more than non-schizophrenics are.

Rokeach eventually showed the three Jesuses an article from a newspaper about the experiment to see what they would make of it. Joe and Clyde found the article amusing, saying that the three guys in the paper were 'completely nuts', but Leon realised the article referred to him and his friends. He became furious and disillusioned with Rokeach because he saw it, understandably, as a betrayal.

But then something remarkable happened. Each of the three Jesuses started to believe independently that the other two were mentally ill and that the kindest thing would be to humour them in their belief. While each man still believed he was the true Jesus, they all concluded that the other two should be allowed to go on believing it out of compassion.

This result was unexpected because the prevailing wisdom of the time was that schizophrenics were incapable of empathy. Schizophrenia and psychopathy were, as far as anyone knew, the same condition. Rokeach's experiment was starting to show otherwise. In fact, as the months went by, the three Jesuses started to become friends and began voluntarily spending time together, even sticking up for each other in arguments with other patients. So although none of them was cured, they did stop talking about their delusions around people who might be upset by them.

This experiment, whatever we might think of it from a moral perspective, showed conclusively that schizophrenia is neither a choice nor something a person can be persuaded out of. It's also not the same thing as being a psychopath. Violent schizophrenics were not lacking compassion; they just didn't see the world objectively. Schizophrenics can't control their beliefs, but they can control how they express them.

As the experiment progressed, the grad students Ron and Dick started to challenge Rokeach and accused him of experimenting maliciously on the three individuals they were starting to care for. They eventually resigned when Rokeach refused to stop the experiment, but thirty years later Rokeach wrote about the trial in his memoirs and expressed deep regret. He admitted that while the three men were not cured of their spiritual delusion, he was cured of his own – his belief in a God-like right to mess with other people's lives.[36]

PART THREE
Surprises

....

'Surprise is the greatest gift which life can grant us.'

Boris Pasternak

'We've found that the world is very surprising.'

John Polkinghorne

'Surpriiiiiise'

Loki, God of Mischief

FOXY BUSINESS

Darwinian evolution is the cornerstone of self-respecting biology, but it deals with processes that stretch back millions of years. This means it's impossible to run experiments to learn *why* a particular trait was favoured, so we often have to rely on speculation to figure out why evolution has done its thing. Not to mention the fact that there's a significant minority of non-scientists who don't even accept Darwinian evolution occurs in the first place.

Most of the time these kinds of debates are arbitrary, but during the 1950s they had a devastating effect on the Russian economy when a biologist named Trofim Lysenko began endorsing the discredited hypothesis that inherited traits are the result of environment rather than mutations in DNA.

To be fair, there are *some* traits which can be – in a very loose sense – environmentally 'passed down'. If a mother is exposed to stress while pregnant, for instance, it can affect the way her daughter's brain develops *in utero*. The daughter can grow up to become an adult with a greater risk of anxiety herself as a result of her prenatal environment . . . which she can pass on to *her* daughter when she gets pregnant, and so on. There's a branch of science called epigenetics which explores how environment impacts gene expression – but make no mistake, it's still genes which run the show. Environment can determine how a DNA strand is decoded, but DNA itself is only changed through Darwinian mutations.

Lysenko believed, however, that you *could* control heritable features in a child by changing the life of the parents. If you chopped the tail off a mouse, for instance, its children were more likely to be born tail-less, and so forth. This idea is not supported by any evidence but it gained significant traction under the dictatorship of Stalin.

There are a number of possible explanations for why this happened. Lysenko was born to a peasant family and the idea of an impoverished child growing up to save Russian agriculture played nicely into the Marxist idea of workers revolting against the middle class. Lysenko's beliefs could also have chimed with the Stalinist drive to engineer people via social control. Or it could have been something as simple as Stalin liking and having a better understanding of Lysenko's approach to evolution.[1]

In any case, the suggestion of genetic inheritance driving evolution was ultimately banned in the USSR, to the point where pro-Darwinian scientists could be arrested for defending the bourgeois facts of genetics.

One notable scientist who lost his job for having Darwinian leanings was Dmitri Belyayev. Belyayev was horrified at the suppression of information taking place in his country and decided to prove that not only did Darwinian evolution work, it did so via genetic mutation and selection.

He teamed up with the ethologist Lyudmila Trut and together they devised a simple yet brilliant experiment. They would get hold of some wild silver foxes, put them in a controlled environment and only allow certain foxes to breed. If Darwin was correct, the trait they chose would become dominant in the cubs, proving it was genetics that drove evolution. And, just to show the far-reaching power of Darwinism, they decided to focus on a behavioural trait rather than a physical one: friendliness.

Their experiment began in 1952 when Trut took 130 silver foxes and set up a breeding farm in Estonia. The foxes were isolated from human interaction to avoid being accidentally trained, and would be observed secretly at a distance for friendliness. Foxes which tried to bite Trut and Belyayev on the rare occasions they approached were not allowed to breed, while foxes which seemed more amenable, were.

The offspring of the first generation were then put through the same selection process, with friendliness being singled out, and so on. If Darwin was right then eventually friendliness would become an inherent feature of the foxes and they would become naturally 'tame' animals without requiring environmental training.

The experiment succeeded perfectly, with the silver foxes becoming friendly and approachable in just ten generations. Trut and Belyayev had shown that evolution worked via genetics, as well as showing that behaviour could be selected by natural processes. But this just confirmed something every sensible biologist already knew. The truly remarkable finding was what else happened to the foxes.

By 1969 Trut and Belyayev had successfully bred into existence a new type of fox, but it wasn't just behaviour which had changed – anatomy had, too. In just seventeen years, the new generation of foxes had floppier ears, curlier tails and rounder muzzles. Their limbs had become stubbier and the colour of their fur had turned patchy. Even their adrenal glands had shrunk.

Without meaning to, Trut and Belyayev had turned wild silver foxes into animals with physical features (and therefore genes) in common with domesticated dogs.[2]

The first implication was obvious. Evolution could happen *extremely* fast. If you could turn a fox into something eerily dog-like in just seventeen years it was reasonable that over a few million years wolves, dingoes, foxes, dogs and coyotes could all have diverged from the same starting creature. Go back tens of millions of years and the canines could have diverged from the bears, pandas and raccoons. Give it a few billion years and Darwinian evolution looks not only reasonable but inevitable.

The second implication is that evolution can, under the right conditions, jump rather than crawl. One of the ideas of early evolutionary theory was that there is a constant rate at which a gene outcompetes its rivals.

This viewpoint, called 'gradualism', is one of the common objections evolution-deniers focus on because it sounds like a solid counter-argument. There are many points in the fossil record which show leaps in the evolutionary process, where lifeforms suddenly go through rapid alterations out of nowhere. This can't happen naturally, say anti-evolutionists, ergo evolution must be wrong.

The Trut and Belyayev experiment showed, however, that it was gradualism which was at fault. It was clearly possible for major changes to occur in short spaces of time, faster than the 'normal pace' of evolution.

This new way of thinking about evolution is called the punctuated equilibrium approach and shows that in an extreme or unusual environment, evolution can actually speed up.

The third implication – and perhaps the most startling – was that selecting for one trait in a species can have an effect on other, seemingly unrelated, ones.

Darwin had originally thought that every feature of an organism is the result of some useful adaptation in the species' past. Evolution doesn't select a gene unless it's helping, so every feature an animal has must somehow be the legacy of a previous genetic battle.

Here's an example of how this might work. Suppose you want to explain why men go bald in later life. We need to come up with an explanation for how balding gave men an advantage, so how about this: having hair on your head protects your skin from sun damage. As men get older they become less appealing to young females and are outcompeted by younger men, but by losing their hair they're more likely to get skin damage. This will encourage younger females to take sympathy on them. This leads to a bonding situation which makes breeding more likely. Sound far-fetched? Well, prior to Trut and Belyayev, there was no alternative to this kind of thinking.

Their experiment put a massive dent in this adaptationist view of evolution. It turned out that evolution doesn't necessarily work by only letting good traits through the Darwinian filter. Sometimes, features of a creature are an unintended by-product of something else.

The human body features at least one hundred thousand different types of protein, but our DNA only contains twenty thousand genes (for contrast, a banana contains thirty-six thousand). Genes have to splice, combine and shuffle around in order to give rise to human complexity, so a mutation in one gene will have a knock-on and unintended effect on other genes.

For example, the 'AR gene' in humans is involved in sperm production and muscle mass, so if your genome mutates to produce more sperm you're also going to become more muscular. But the AR gene is also involved (by coincidence) in hair growth. Mutating the gene in order to give better sperm production will also affect how much hair grows, i.e.

baldness in men may not be an advantage at all, it may simply be the price a male human pays for making more sperm cells.[3]

In the silver fox experiment, the scientists were selecting for friendliness, but the underlying gene they were giving advantage to was also in charge of bone and fur growth. Trut and Belyayev managed to show that:

- Behaviour is influenced by genetics
- Evolution can work at breakneck speed
- Evolution can change speed
- Multiple features can be changed at once
- Not every adaptation is beneficial

This is perhaps the most profound experiment of twentieth-century biology (as well as the cutest).

Rat race

In 1947 John Calhoun asked his neighbour if he could build an enclosure in his garden for a science experiment. His neighbour agreed, not realising what he was saying yes to, because Calhoun immediately fenced off a quarter-acre enclosure (about four tennis courts) and populated it with pregnant female rats.

What he wanted to know was whether rats had the ability to exercise sexual self-control. In terms of physical space, an enclosure of that size could house five thousand rats easily, but what would happen to their numbers as they reached this limit? Would they know to stop having sex?

Normally a population is limited by things such as food and water, so Calhoun decided to keep those in plentiful supply to make sure the rats could keep breeding until they were shoulder to shoulder. But they didn't fall for it. Calhoun found that the rat population peaked at 150 and stayed there. Somehow the rats *were* able to sense that they were boxed in and although they had enough nutritional resources, the limited space curbed their sex drives.

By 1954, Calhoun was working at the National Institute of Mental Health in Maryland and decided to recreate his earlier experiment, this time in a 10 × 14-foot enclosure he dubbed Rat Universe 1. Again, the population got to a steady number and stopped as if the rats collectively knew they were increasing their population density.

Over the next two decades, Calhoun repeated his experiment with minor variations all the way up to Rat Universe 24, each time with the same result: a rat population will level off in an enclosed space no matter what resources are provided.

So, for Rat Universe 25 he decided to push his experiment to the extreme: he built a city for his rodents (deciding to use mice instead because they were smaller). He constructed a miniature metropolis of sixteen 5-foot-high skyscrapers, with apartments inside, along with a town square for socialising, an ambient temperature, a constant food supply, fresh water, material for nests and absolutely no diseases. Mouse city was a utopia.

The population initially doubled every two months as per normal until there were over two thousand mice living in his city, at which point it started to level off. This was, by now, an expected result – but then the changes crept in. Disturbing ones.

Violence started to break out amid the mice. This was followed by a period of intense hyper-sexuality that eventually devolved into an unending period of apathetic asexuality, during which most of the inhabitants stopped trying to mate altogether.

It was just about conceivable that the mice were able to sense overcrowding, which made them breed like crazy (perhaps with the urge to continue their genes before the sex stopped). But something else happened which couldn't be explained – the pregnancy rate started failing. Dramatically: 96 per cent dramatically. Not only that, mothers started attacking their children and cannibalism became a way of life as Mouse City descended into a putrid dystopian chaos of misery, cannibalism and regret. Basically it was like Leeds.

The majority of Calhoun's mice ended up becoming docile and emotionless, sitting in the city's town square like zombies, waiting for food to arrive, with occasional fights breaking out.

More alarming was that the biggest males staked out large territories in the skyscrapers, sequestering a number of apartments as their domain. These gang-leader mice would also keep a harem of females on site who spent their time lounging around, occasionally grooming but otherwise moving only to have sex with the dominant male.[4]

Parallels were immediately drawn with human cities. Apparently overcrowding could lead to a breakdown in pro-social behaviour, an increase in apathy, an increase in infant mortality and the rise of polygamy. Most interestingly though, it was the mice who were already dominant before the experiment started that ended up displaying the gang-leader behaviour.

What Calhoun's experiment had shown was that overcrowding leads to stress which magnifies all sorts of unpleasant things. Cram enough living things into a tight area and you're not going to see a utopian equilibrium. To some, his experiment was a stark warning of what can happen if you crowd a population of humans into a city. To others it bore no significant parallels to humans, however, because (simply put) people aren't mice.

It's worth comparing Calhoun's experiment with that of Bruce Alexander who ran a similar test to see how environment affected behaviour. In 1978 Alexander conducted trials in which rats were given the option of consuming water laced with morphine vs sweetened water. He put one group of mice in confined cages typical for a lab-rat, but for another group he created a rat-topia.

He called it Rat Park and it was designed to offer not only necessities but luxuries too. Rat Park had food, water, heat lamps, nesting material, ample space (two hundred times more than a normal cage), plus balls to play with and decorations on the walls. He found that rats who lived in Rat Park did not develop dependencies on morphine to the same degree as those in a standard lab cage.

The experiment was designed to show that since rats were used so regularly in scientific studies, the very act of putting them in a cage was likely to cause atypical behaviour.[5] But it also ended up showing that there could be a link between susceptibility to drug addiction and environment.

Helpful bugs

Schizophrenia is a devastating condition in which the patient is unable to distinguish reality from fiction and has difficulty putting information into logical sequences in their head. In its most serious form it affects 1 in 300 people and can utterly incapacitate its sufferers when left untreated.

We still don't know what causes it, but much of what we do know comes from a series of experiments involving drugs, spiders and a lot of urine. Unfortunately the story revolves around three scientists named Peter, Hans-Peter and Hans Peters, so this might get confusing. Here goes.

The first Peter was Peter Witt, born in 1918, who by 1944 had become Germany's leading expert on drug pharmacology. After the Second World War he was even recruited by the American military to analyse Hitler's medical records and see if he had been on drugs during his bunker-suicide phase (he was – over one hundred a day).[6]

Witt's life took an odd turn when he was approached by Hans Peters, a colleague at Tübingen University who was studying spiders. Hans Peters wanted to observe spiders spinning their webs but was frustrated by the fact that spiders inconsiderately do it in the middle of the night. He asked Witt if it was possible to give spiders stimulants or sleeping medication to regulate their cycles to a daylight-friendly regime. Witt came up with the idea of dissolving drugs in sugar-water and feeding it to the spiders to see what would happen.

He tried giving the spiders strychnine, morphine and methamphetamine, but the experiment didn't change their routine – Hans Peters would have to live with getting up early. But an unexpected result caught Peter Witt's interest. When the spiders had been given drugs, it changed the shape of their webs.

Normally it's difficult to assess the effects of a drug precisely because animals don't always do things which are mathematically measurable. But because you can quantify the features of a web easily, e.g. angles, crossovers, distance between strands, etc., it gave Witt a chance to numerically measure drug impact on spider nervous systems.[7]

By 1952 he had settled at the Raleigh Mental Health research department of North Carolina and spent a decade giving drugs to orb web spiders, including mescaline, magic mushrooms, cocaine, caffeine, Valium™, marijuana and LSD, although he failed to give them alcohol because the spiders were not interested in the taste.

Witt concluded that while his spiders 'loved the drugs' it caused major alterations in their ability to function and web-spin usefully.[8] This might have seemed like a useless area of scientific research until, completely out of the blue, his work was used to settle a major debate about the cause of schizophrenia.

One of the proposed explanations at the time was that the illness was the result of a hallucinogenic chemical produced in the brain, i.e. a schizophrenic person was naturally producing their own internal form of LSD.

A drug specialist named Hans-Peter Rieder decided to find out if schizophrenics really *did* have hallucinogenic chemicals floating around their body and used Witt's technique. Witt had shown that the presence of hallucinogens can be detected from spider-web patterns, so Hans-Peter collected 50 litres of schizophrenic urine and fed it to spiders to see if their webs changed.

He found that while spiders don't particularly like the taste of human piss, there was no difference in their web architecture, meaning schizophrenia was not caused by a self-generated hallucinogen.[9]

Unhelpful bugs

When something causes damage to the body, pain sensors called noci-ceptors send a signal to the brain via C-fibre neurons (dull, aching pain) or A-fibre neurons (sharp, focused pain). People have tried to classify pain numerically for a long time, but the most widely used pain scale was put forward unintentionally by an American entomologist named Justin Schmidt.

In 1984, along with his collaborators Murray Blum and William Overal, Schmidt was studying hemolysins – chemicals found in the venom of wasps, bees and ants. Hemolysins break down blood cells, so Schmidt wanted to know if there was a link between how much hemolysin a bug's venom contained vs how toxic it was. It turned out there wasn't any such link, but something else in his paper caught everyone's attention. In the process of milking the various insects, Schmidt received a lot of stings and noticed that sting pain is somewhat uniform. In fact, all the stings he received fell into only four broad categories.

He decided to rate them on a scale of 0 to 4 which he included in his finished paper as a sidenote.[10] The Schmidt pain scale (along with his personal descriptions of what they feel like) runs as follows:

0 Virtually no pain – the stingers don't get through the skin (giant sweat bee, cuckoo bee, club-horned wasp).
1 Light, minor, almost trivial pain – comparable to a mild static shock (red fire ant, tropical fire ant, southern fire ant).
2 Painful – like someone putting out a cigar on your tongue (honeybees, glorious velvet ant, large tropical black ant, western yellow-jacket).

3 Sharply and seriously painful – like having a rat trap slam shut on your fingernail (ringed paper wasp, cow-killer velvet ant, harvester ant).

4 Traumatically painful.

The only creature to earn a score of 4 was *Paraponera clavata*, the South American bullet ant. The pain was described by Schmidt as 'pure, intense, brilliant pain, like walking over flaming charcoal with a 3-inch nail embedded in your heel'.[11] It lasted uninterrupted for four hours before dulling to 'extremely painful' for another twenty-four hours, during which time Schmidt suffered numbness, trembling, hallucinations and began crapping blood.

Schmidt's pain scale was very simple, but this turned out to be its strength. While other pain scales existed, they ran into trouble by being too precise. How can you distinguish a pain rating 3.4 from a pain of 3.5, for example?

The Schmidt scale simplified the problem with four easily identifiable categories that everyone seems to agree on. Precise measurements of pain differ from person to person, but everyone apparently feels pain in roughly four ways. The Schmidt pain index caught on, and the following year a scientist named Christopher Starr outlined a method for standardising the score.

Firstly, 'reports should be made only by adult observers in good health'; secondly 'a ranking should never be based on just one sting'; and most importantly 'reports on stings received through free attack by the insect are preferable to those deliberately induced', although he did admit 'we are not always so fortunate, as to be attacked by those species of special interest', and that sometimes a sting would have to be forced.[12]

Since inventing the scale, Schmidt has encountered only four species whose pain reaches a '4'. The warrior wasp (which stung him on the forehead), two variations of tarantula hawk wasp (*Pepsis grossa* and *Pepsis thisbe*, which stung him multiple times) and the bullet ant, which he still considers the worst.

Bullet ants are used in the coming-of-age rituals of the Sateré-Mawé tribe in the Brazilian rainforest, where a young man must put his hands into a glove that has had ants sewn into the lining (stingers inward). The man must wear the glove for five minutes over his head and must complete this ritual twenty times before he is considered mature.

Schmidt has categorised over ninety-six species of insect sting and endured over one thousand injuries, which he catalogues in his book *The Sting of the Wild*.[13] Fortunately, most of these stings were 'scored in the field' but 'in exceptional situations where normal stings were not received . . . intentional stings were received by forcing the insect to sting the side of the forearm'.[14]

SHOCKING

The wires our body uses to convey sensory information to the brain are called neurons, but neurons are also used in reverse to send signals from the brain (itself composed of neurons) to the muscles for movement. The discovery that all movement is caused by electrical signals zapping into the meat of our flesh was made accidentally in the 1790s by the Italian physicist Luigi Galvani.

Galvani was, for reasons it's best not to dwell on, hanging frogs' legs on brass hooks, suspending them from an iron railing in his garden and seeing what would happen when lightning struck them. (The 1790s were a great time to be alive.)

Galvani and his assistant were working with a scalpel that had acquired an electric charge from another experiment involving a Leyden jar (an early form of TASER™). While holding the scalpel, Galvani's assistant's hand casually brushed the iron railing from which his frogs' legs hung and they discovered, to their delight, that the legs began twitching.[15]

This was not the result of lightning and, more importantly, the electric charge hadn't touched the amphibious limbs directly. Apparently, an electric shock to the railing connected to the hooks, allowing a current to pass into the legs and animate them from a distance.

What Galvani's assistant had discovered was that electrically induced muscle spasms were not merely a side-effect of lightning – they were the very method by which the body created movement. Galvani proposed the idea that muscles contained a fluid which moved through them (called animal electricity) and although he wasn't correct, it was moving in the right direction. Thanks to the work of another great Italian researcher, Alessandro Volta, we eventually got to the correct explanation a few years later.

The head of a neuron is full of salt water and when holes in its
membrane open and close, it changes the concentration. Since salt particles have tiny charges, changing their concentration will cause charge
imbalances on either side of the neuron's wall. We call this charge imbal-
ance between two places a 'voltage'. Neuron voltages are thus controlled
by letting squirts of salt water into and out of the neuron which can then
be conducted down their length, allowing us to control whichever muscle
we need.

Another Italian physicist named Giovanni Aldini took Galvani's
discovery further in 1803 and used electric shocks to animate the corpse
of executed murderer George Forster. Aldini showed that not only frogs
but humans too owe their animation to electricity. There is nothing
spooky about our ability to move: it's simple biophysics.[16]

These experiments were a regular topic of discussion with the poet
Lord Byron and his friends – which included a young Mary Shelley.[17] In
1815, while on holiday in Germany, staying a few miles from Frankenstein
Castle in the Odenwald mountains, Lord Byron proposed a writing
competition among the group to come up with a work of horror fiction
around the idea of reanimating the body with electricity. Fairly obviously,
Shelley won.

WHEN YOUR NUMBERS ARE OFF

Scientists can be quite obsessive. The English chemist John Dalton, for instance (who discovered colour-blindness in the late eighteenth century), had an obsession with swamp gas and would go daily to his local bog to take readings on consistencies and temperatures of the emissions. Dalton also kept meticulous diaries and records on weather patterns for forty years, which gave us some of the earliest and most thorough data we have on meteorology.[18]

My personal favourite number-obsessed scientist, however, was the brilliantly named Sanctorius Sanctorius who lived in Italy at the turn of the sixteenth century. Sanctorius was a friend of Galileo and a well-respected physician to the aristocrats of Venice. Among other things, Sanctorius is credited with having invented the pulse meter, the wind gauge, the thermometer and a somewhat less universally loved invention which can only be described as a set of 'hanging toilet-scales'.

Sanctorius built a set of giant portable scales resembling a swing-seat, which would tell him his weight in a variety of circumstances. He would sit in these dangling scales while working, while exercising, while having sex and while both eating and excreting. Which he did for thirty years (by which I mean, he kept the records for thirty years, not that he was sitting on the toilet for half his life).

Sanctorius was mostly doing this for fun (I assume), but as he studied the rise and fall of his body mass, he noticed something odd – the mass of food going into his body was always greater than the mass of stuff coming out. In fact, 63 per cent of his food never made a reappearance at the other end, meaning the food was somehow being absorbed. Sanctorius's fanatical attention to detail gave us the first unexpected clue that food gets digested and incorporated into the body.[19] But the most famous accidental numbers-discovery in science didn't relate to Sanctorius's anus, but to Uranus.

YES, I'M PROUD OF MYSELF

In 1821 the French astronomer Alexis Bouvard published a 110-page book with the gripping title: *Astronomical Tables*. What was this book? Put simply, it was 110 pages of endless numbers predicting the positions of planets in the solar system based on readings Bouvard had taken. That's it. (For important comparison my own book *Astronomical*, which you can buy at all good book shops, has funny diagrams and references to Pringles™!)

Bouvard's book begins with a twenty-eight-page introduction (seventeen of which are more tables) outlining his methods of calculation, before he barrels ahead with nothing but numbers calculating the orbits of Saturn, Jupiter and Uranus.[20]

It took him thirteen years to complete and you might assume that most people would raise an eyebrow at this achievement before politely nodding and saying 'Sure Alexis, I'll definitely read it later.' But astonishingly people actually sat down and read the thing from cover to cover. Not only that, they also began comparing it with their own results to see if they matched up. It was through these comparisons that Bouvard noticed something disturbing. His calculations for Uranus were wrong.

Bouvard had taken Newton's laws of gravity and applied them to the outer planets, predicting where they should be in the sky. But the Uranus numbers were off. It wasn't always orbiting where his equations predicted, which could have been extremely embarrassing. To have spent over a decade calculating something only to find you've forgotten to carry over the 1 is a nightmare.

After double-, triple- and quadruple-checking, Bouvard decided that he hadn't made a mistake. There *was* an anomaly, but it wasn't with his equations. It was with the planet itself. In science, you always throw out your theory when the data says there's something wrong, which meant one of two things was happening. Either Newton's theory of gravity was incorrect (not bloody likely) or there was something pulling Uranus out of its normal orbit.

Bouvard decided there had to be another planet further out, whose gravity was yanking things off course. He went to his grave convinced there was an eighth planet, but tragically it wasn't until three years after he died that the German astronomer Johann Galle confirmed the existence of Neptune.[21]

The idea of poring through a 110-page book of nothing but numbers might seem daunting and tedious, but consider that the Large Hadron Collider at CERN in Switzerland generates one petabyte of data – the equivalent of ten thousand 4K movies – per second. There's just no way of reviewing that information, so the LHC computers simply delete 99.996 per cent of it.[22]

There are nifty algorithms which try to search for interesting results, but there's no other way of saying it: we only ever see 0.004 per cent of the data the LHC gathers. Could there be discoveries hidden in the rejected data we don't have time for? Absolutely. But you might as well shove those numbers up Uranus for all the good they'll do.

Newton's unlikely rainbow

Most people don't manage to write a masterpiece. Isaac Newton wrote two, however, because he had a stubborn habit of being the cleverest person in the world all the time.

The first of his *magnum opuses* (even typing the plural of *magnum opus* seems ridiculous) published in 1687, was *Principia Mathematica*, which defined the laws of motion, gravity and ultimately physics itself. His second *magnum opus*, published in 1704, was *Opticks*, which defined the physics of light.

Within its pages, Newton gave the first successful and accurate account of where rainbows come from, something nobody else had come close to doing. It was already known that you could take sunlight and turn it into a spectrum by forcing it through a prism, but the understanding pre-Newton was that a prism was adding colours to the beam. Newton discovered this was the complete reverse of what was going on.

He found that if you took a rainbow and passed it through a second, upside-down prism, you got white light back out. If a prism added colour to a beam, there was no way of explaining how a second one was removing it. The answer had to be something else.

The physics of rainbows is as complicated as rainbows are beautiful and I've devoted a lengthy appendix to them (Appendix 1), but the key takeaway is that different colours are the result of the same basic phenomenon. Red light isn't made of a different thing from violet light: they're just different ends of a spectrum of values.

Newton decided to split the range of possible colours into seven categories for no reason other than he thought seven was a magical number. There seemed to be only six colours, though: red, orange, yellow, green, blue and violet, so he invented the term indico (later indigo) which, let's be honest, is just dark blue.[23] Newton didn't know what light was made of, however; that understanding arrived much later and happened as the result of eight separate accidents.

Accident One

In 1820 the physics lecturer (and brother of the Danish prime minister) Hans Ørsted was preparing to show his audience the wonders of a new device taking Europe by storm: the battery. As he was getting ready to deliver his talk, he placed his compass on the desk beside the battery and saw, to his confusion, that the compass needle was no longer pointing North. Instead, it was pointing parallel to the battery. He had discovered that electrical currents generate magnetic fields.[24]

A few years later, the British scientist Michael Faraday made the opposite discovery – that moving a magnetic field near a wire will generate an electric current. Electric charge and magnetic attraction were connected.

Accident Two

The Scottish scientist James Clerk Maxwell was a huge fan of Michael Faraday and wanted to establish a unifying theory of how electricity and magnetism are linked.

In 1865, two years before Faraday died, Maxwell proposed that there was an invisible force field around us (like gravity) which can be agitated by magnets or particles with charge. Since both magnets and charges feel this field simultaneously, you can't affect one without affecting the other. Disturb the field by waving a magnet about and you'll create ripples which are felt by a nearby electric charge, and vice versa.

Maxwell was able to use the measurements of Ørsted and Faraday to calculate the speed at which waves would move through this 'electromagnetic' field, and this turned out to match the speed of light (300 million metres per second). Without meaning to, Maxwell had discovered that light was made of ripples in the EM field.[25]

Most of the time, the EM field is dormant and we aren't even aware that it's around us. But when something causes it to oscillate, the vibrations can spread like ripples across a pond. Our eyes perceive these field disturbances as light. Different colours are down to the size of the ripple, e.g. red light is a broad ripple in the EM field while violet light is a narrow ripple. Which leads to a great question: can we make waves wider than red or narrower than violet? Are there colours our eyes simply aren't able to see? Turns out the answer is yes.

Accident Three

The British astronomer William Herschel (discoverer of Uranus) wanted to know which colour contained the most heat so in 1800 he set up a simple experiment: split sunlight into its components and put a thermometer in the path of each colour to measure the temperature. Herschel was an excellent lab scientist, though, and he knew that it would also be a good idea to put an eighth thermometer beside them to get a baseline reading of the room.

He put his additional thermometer beside the red light and waited for the temperatures to change, but when he checked the readings they didn't make sense. The colours of light seemed to be much colder than the room itself. The one he had placed by the red beam was far warmer.

Herschel quickly realised his room temperature thermometer was the one throwing him off, over-reading the room's temperature dramatically. When he placed it in other locations he got a more sensible value, but when he put it next to the red beam, its temperature rose. He had discovered an extra colour of light, which he called *infra*-red from the Latin 'below red'.[26]

Accidents Four and Five

The following year, German chemist Johann Ritter read about Herschel's work and tried to measure the opposite effect – a cooling ray beyond violet. His experiment failed and the thermometer did not record a temperature increase, but while carrying out the experiment, he noticed that the photographic plates he had stacked behind his thermometer were developing faster than they would in normal light.

There apparently *was* a colour beyond violet, but one which interacted with photography plates rather than thermometers.[27] It was eventually named *ultra*-violet from the Latin 'beyond violet'.

To understand this perplexing result we need to know how photographic plates work in the first place (they too were discovered by accident). Back in 1717 the Swiss chemist Johann Schulze had been experimenting with a solution of silver nitrate and chalk. One afternoon he absentmindedly placed the bottle on his windowsill and came back later to find the contents had turned a dark grey except for one area, where a white line had formed in the liquid.

When Schulze looked out of the window he saw a washing line strung up at the same angle as the trail inside the bottle. After a bit more experimentation, Schulze discovered that sunlight had the ability to turn silver nitrate into a dark powder, which made the chalk in the solution stand out by contrast. The washing line had blocked the sunlight from hitting parts of the bottle and so part of the contents stayed white, like a reverse shadow made of liquid.[28]

The chemistry going on inside the bottle was subtle. Silver nitrate is composed of charged silver particles floating around in the presence of oppositely charged nitrate particles. The silver and the nitrate are attracted to each other, which keeps everything stable, but when a high-energy beam strikes the particles it provokes a change.

The sunlight allows the silver particles to lose their charges and bind to each other (see Appendix 2 for more details), forming solid silver dust. Silver dust is greyish, which means when you shine a beam of high-energy light at silver nitrate, wherever the beam touches will form a dark residue.

The French scientist Joseph Niépce took things one step further, coating a board with silver nitrate and leaving it on his windowsill in the hope of creating an imprint of the view. His experiment was a success and he managed to take the very first photograph: 'View from the Window At Le Gras'. The entire photography industry thus owes its existence to Schulze leaving a bottle out in the sun.

Fast forward to Johann Ritter's lab, and his observation of UV light now makes more sense. Visible light contains enough energy to discharge silver nitrate particles in a photography plate, but ultra-violet has even more energy, causing the reaction to take place faster.

Ritter had also made another important discovery: light energy is not as simple as 'hot at one end, cold at the other'. What's really going on is that the size of a light wave has to correspond to the size of the thing it's interacting with.

The size of an infra-red wave corresponds to the size of atoms and molecules, meaning they will vibrate when infra-red is shone on them. This is what we perceive as temperature: the jiggling of atoms and molecules.

Visible light waves are too narrow, however, and pass right through the atoms without disturbing them. Even though visible light carries more energy than infra-red, the beams aren't able to move the atoms because they're the wrong size. What they *do* interact with are electrons going around the outsides of atoms.* Smaller wavelength = smaller thing gets moved.

Johann Ritter is also known for having carried out experiments with batteries to test the effects of electrocuting his eyeballs (which caused hallucinations), his nose (which caused sneezing) and eventually his penis (which caused a scandal). Ritter even went so far as to declare that he had fallen in love with his battery and planned to marry it.[29]

---

* Strictly speaking, visible light waves aren't the same size as the electrons themselves. Electrons orbit atoms on shells at specific distances from the nucleus. It's the jumps *between* these shells which correspond to the size of visible light waves. A visible light wave (or an ultra-violet one) will thus cause electrons to jump between these shells in a process called 'quantum leaping'.

Accident Six (sort of)

With the knowledge that the spectrum of electromagnetic waves aka 'light' didn't stop at red or violet, other invisible colours were soon sought. The colour below infra-red was deliberately generated in 1894 by the Indian physicist Jagadish Bose and is called microwave. But although microwave light wasn't discovered by accident, its most famous use *was*.

In 1945 the American physicist Percy Spencer was working with military grade microwave emitters to see if they could be used to bounce off enemy aircraft. While working with one of these emitters, Percy noticed one of two things, depending on which source you read.

According to one version of the story he noticed a candy bar in his pocket melting (allegedly a Mr Goodbar™),[30] while other sources state he noticed his own body close to the emitter getting warmer.[31] Whatever it was, Spencer discovered that microwaves being given off by the scanners were vibrating fat and water molecules in everything around them, raising their temperature and cooking them from the inside out.

He tried heating a bowl of popcorn by holding it up to the emitter, followed by an egg (which exploded in his assistant's face) and realised he had discovered a way to heat any material containing fat or water . . . which is almost all food. Once the war was over, Spencer repurposed the microwave emitters he had been working on and sold them as the first microwave ovens.

Accident Seven

At the other end of the EM spectrum, the colour beyond ultra-violet was also discovered by accident, this time by the German physicist Wilhelm Röntgen who was experimenting with a cathode-ray lamp. Cathode rays are high-energy beams of electrons and Röntgen wanted to see how they would impact fluorescent paint.

On 8 November 1895, he was setting up his experiment and decided to put his cathode lamp inside a black cardboard tube so he could block the light coming out in all directions and focus the beam to a thin circle.

But as he was getting ready, he noticed the fluorescent screen against his office wall starting to glow. The cathode lamp was pointing away from the screen, so there was no light source to activate the chemicals. Unless, he thought, there was a beam of light coming off the lamp which was so powerful it could pierce the black cardboard.

After further experimenting, Röntgen selflessly volunteered his wife Anna to stand in front of the beam and found that by placing a photographic plate behind her, the beams would pierce her flesh but not her bones, leaving an imprint of her skeleton. Upon seeing the outline, Mrs Röntgen supposedly remarked 'I have seen my own death'.[32]

Röntgen called these mysterious beams 'X-rays' as a placeholder name (X being a common symbol to denote the unknown quantity in an equation). He intended to come up with a better name later, but 'X-rays' was so catchy it stuck.

Accident Eight

As we've already seen from Johann Ritter discovering UV rays, silver nitrate plates will react to high-energy beams, and they did it again in 1896 when the French physicist Henri Becquerel wanted to try recreating Niépce's photographic experiment.

He prepared his silver nitrate plates to capture an image through his window but unfortunately it was a cloudy day, so he stuck the plates in his drawer over the weekend and went home.

When he came back on Monday, something impossible had happened. The photographic plates had somehow been developed. He had placed them underneath his Maltese Cross war medal and, without any light source inside the drawer, the plates had captured a silhouette of the medal.

Becquerel reached out to his friends, husband and wife duo Marie and Pierre Curie, for their advice, and with a little sleuthing the Curies deduced what was happening. There was one other item in the drawer besides the plates and the medal – a bottle of uranium sulfate which the Curies found was spitting out high-energy beams. These beams were activating the silver particles in the plate the same way sunlight would, but the medal was made of thick metal and blocked their path.[33]

Marie Curie named this newly discovered phenomenon radioactivity, from the Latin for wheel spoke (*radius*) and the Greek for beam (*aktinos*), since these beams spread out like spokes on a wheel. They had discovered the next colour of light beyond X-rays: gamma rays.* These rays are about the size of an atomic nucleus and pack such a punch they can vibrate the very atomic cores of your body.

Once the Curies had discovered there was a new type of beam to be studied, they began buying mountains of pitchblende, a mineral known to contain trace amounts of uranium. Through studying this ore they learned that heavy atoms tend to be the ones which gave out radioactive emissions (see Appendix 3), and they also discovered two elements in the process as a bonus.

---

* It's a tiny bit more complicated because there are four different types of radioactivity. Gamma rays are the next colour of light beyond X-rays, but heavy atoms also emit high-energy particles which have radioactive properties. The Curies didn't have the technology to distinguish high-energy beams from high-energy particles so they referred to them all as 'radioactivity'. It was technically Paul Villard who, in 1900, discovered that the radioactive emissions from uranium were EM beams, which *he* named gamma rays.

Searching in Ernest

In 1908 Ernest Rutherford won the Nobel Prize for Chemistry. He had built on the work of the Curies and discovered there were different types of radioactivity emitted from large atoms. Not only did heavy elements give out high-energy beams of gamma light, they would also eject tiny 'bullets' of matter too. He had no idea what was causing these ejections but it was an undeniable phenomenon. He called the tiny bits of atomic debris 'alpha particles'.

Many other scientists would have rested on their laurels at this point and enjoyed eating for free in any scientist-heavy restaurant in the world, but Rutherford was keen to push his knowledge deeper. He wanted to know where the bullets were coming from: but in order to do that he'd have to probe the inner structure of the atom.

Atoms are so small you can't just look at them under a microscope because visible light is too big. You could line up a string of atoms for every human being on the planet and it would barely reach 8 centimetres. If you want to study their behaviour you need to get creative, which Rutherford definitely was.

The son of a New Zealand sheep farmer, Rutherford had a reputation for unconventional experiments nobody else would think to try. He liked to prioritise left-field approaches over money and believed fancy equipment and funding made scientists lazy.

The understanding of the atom at the time was to think of it as a cloud with sub-atomic particles called electrons dotted about within it. Rutherford's former supervisor, J. J. Thomson, had already discovered these electrons, and everyone imagined the atom as being like a plum pudding.

What was difficult to explain was that the alpha particles which came firing out of atoms had the opposite electric charge to electrons, so where were they coming from? Was the atomic cloud breaking down? Were there other particles mixed in among the electrons?

To get a look, Rutherford hired the German engineer Hans Geiger to build a device which would react when particles collided with it (the famous crackling Geiger counter) and set it up behind a thin sheet of gold atoms. On the other side of the gold sheet he placed a known alpha-particle emitter, radium, the first element the Curies discovered. The radium would fire its alpha particles through the gold sheet and the detector on the other side would show where they ended up, creating a sort of atomic splatter pattern.

By leaving it running for several hours, he could get a display of where the alpha particles were arriving, which would tell him something about the density of the atoms in the gold sheet. If the atomic cloud was thick and sponge-like, for instance, the alpha particles would be deflected at a wider angle than if it were light and fluffy.

Geiger put his PhD student Ernest Marsden in charge of the experiments, and it was he who discovered something none of them anticipated. While measuring the angle of deflection for each alpha particle, Marsden found a particle which had been deflected by more than 90 degrees. A deflection of two or three degrees was expected, 10 degrees was just about conceivable, but 90 was impossible because the alpha particle would have to be making a right-angled turn when it travelled through the gold. To be deflected by *more* than 90 degrees meant it was hitting something inside the atom and bouncing back.

And it happened again. And again. And again. Marsden found that roughly one in ten thousand alpha particles were bouncing back toward the radium source,[34] which Rutherford described as like firing a 15-inch cannon shell at a piece of tissue paper and have it bounce off.[35]

Whether this experiment counts as accidental is murky, because Rutherford had obviously suggested they try putting the detector at wide angles to the gold sheet. Did Marsden maybe misunderstand him and put the detectors on the same side by mistake? It's hard to say, but the discovery meant everybody had the atom wrong.

The conclusive solution to the puzzle of atomic structure was eventually presented by the Danish physicist Niels Bohr. Bohr was a scientist and footballer (he played in goal for Denmark at the 1908 Olympics[36]) who hit on the idea that atoms must have small, dense nuclei at their centres with electrons orbiting them at fixed distances, like planets round a sun.

The Nobel Prize that wasn't

In 1934, Enrico Fermi was heading up a particle physics lab near the Panisperna monastery in Rome. After Rutherford's discovery that atoms had a lump at the centre, other scientists had identified that it was made up of two types of particle: protons and neutrons.

It was also known that while protons followed a predictable number sequence, neutrons did not. There was an element with one proton (hydrogen), an element with two (helium), one with three (lithium) and so on, but neutrons didn't follow any pattern. It was possible to get different types of lithium (all with three protons in the centre) behaving the same way in chemical reactions – just with different numbers of neutrons, giving the atoms different masses.

What was also known was that when you got to the 92nd element, uranium, the atoms didn't get any bigger. Nature, for some reason, didn't go above atom 92. Fermi and 'The Panisperna Boys' (as they were known) decided to try and beat Nature at her own game. They were going to go a step further and make an artificial element, one with 93 protons. But how do you add a proton to an atom's nucleus and turn it into a new element? This was alchemy.

The tantalising answer came from the American scientist James Chadwick who proposed something clever. Certain elements were known to eject neutrons from their nuclei, especially big atoms like polonium. They had so many particles in their cores, jostling for position, that they would sometimes become unstable and eject a neutron to make room.

This ejected neutron moved so fast it normally shot through everything it came into contact with (this was the problem at Chernobyl in April 1986 where radioactive minerals used in the reactor were emitting high-energy neutrons). But Chadwick suggested you might slow the neutrons down by passing them through a dense material first. When they emerged, they would be moving more slowly and wouldn't fly through your target atom – they might even become lodged inside its nucleus.

The nucleus that absorbed this neutron would now have to rearrange itself; and if you were very lucky, some of the nuclear particles would swap identities.

Neutrons and protons can turn into each other in a process we do not understand. We just know that once in a while a neutron, if disturbed, can turn into a proton. So if Fermi and his team could fire a slow neutron at a uranium atom, the nucleus might absorb it, rearrange itself and by chance turn into element 93.

Since this process would be rare, however, proving it happened would require analysis of something called atomic half-life.

Radioactive atoms are all on the verge of spitting out particles, but you can never predict exactly which atom is going to decay. Instead, you have to look at a group of them and make a general statement about how long, on average, it should take for half of them to break down. It's a paradoxical result because while it's not possible to predict the behaviour of a single particle, you can reliably predict the average behaviour of millions of them.

Thus, while you can never say which atom is going to undergo radioactive emission, you *can* predict reliably how long it will take before 50 per cent of them have done so. This is the 'half-life' of that element; the time it takes for half the atoms in a sample to decay. And the useful thing about half-life is that every atom has its own unique half-life number.

So when Fermi looked at the results of his experiment in 1934 and saw he'd got an atom with a half-life of six hours he was on the edge of his seat with excitement. No element had a half-life with that value. This was something entirely new to the Universe. He had apparently succeeded in creating a new type of atom.[37]

He published his results, claiming to have found radioactive signatures for elements 93 and 94 which he named ausonium (after the modern Greek for Italy – *Ausonia*) and hesperium (after the ancient Greek for Italy – *Hesperia*).

Fermi had done the impossible and was hailed as a hero. In just four years he was awarded the Nobel Prize, not only for the discovery of two new elements but for their creation as well. There was just one teensy tiny, ever-so-minor, inconvenient problem. Fermi was completely wrong.

He was still one of the greatest particle physicists of all time and absolutely deserved Nobel recognition, no denying that. It's just that the thing he got the prize for, technically speaking ... I mean *really technically speaking* ... he hadn't actually done. Because there was another explanation for the radioactive readings he detected. But this explanation was even stranger than making a new atom, so nobody took it seriously.

Three months after he published his alchemical claims, the German chemist Ida Noddack published a response saying she wasn't convinced. A chemist of considerable renown in her own right, Noddack said there was another possibility which had to be ruled out first.

What if the slow neutron Fermi had fired at the uranium atom had destabilised it so much that it cracked in half? The smaller atoms left over would have an unexpected and random arrangement of neutrons in their nuclei which would lead to half-life signatures unlike anything found in nature.[38]

Noddack's idea was ignored because the idea of splitting an atom in half was even more extreme than creating a new one. The very word 'atom' came from the Greek for 'unsplittable'. She couldn't be right. Except she was.

After some investigations by Lise Meitner, Otto Hahn and Irene Curie (Marie's daughter) it turned out that Fermi and his team had accidentally created a new version of the element technetium (element 43). The version they created had a unique half-life which had never been recorded so Fermi and his boys had not created a new, bigger type of atom – they had split an already existing one.[39]

Otto Hahn was awarded the Nobel Prize for Chemistry in 1944 for proving atoms could be split (Lise Meitner was infuriatingly overlooked), but it was actually Fermi who achieved it first without even realising. Fermi's Nobel Prize is unique therefore, having been awarded by mistake, but still being deserved because by coincidence he achieved something which would have won him the prize anyway.

The real creation of element 93 was achieved in 1940 by American physicist Edwin McMillan, who named it neptunium (Neptune comes after Uranus) and he was awarded his own Nobel Prize in 1944. The second person to win the Nobel Prize for making the 93rd atom.

WHO ORDERED THAT?

By the 1930s physicists had nailed down all the key particles that had to exist. Everything was made from just four things: protons, neutrons, electrons and a less common (but very important) particle called a positron. With those four ingredients we could account for all known matter in the Universe.

The positron had been discovered by American physicist Carl Anderson in 1932 using a crude but effective device called a cloud chamber, which is great fun to play with (I built one once because I do things like that). You flood a small tank with a volatile vapour and cool it to the point where the vapour is on the boundary between gas and liquid (rubbing alcohol for vapour and dry ice for coolant works well).

Any agitation to this vapour, like a high-energy particle moving through the tank, will cause the vapour to condense around it, forming a tiny cloud-trail for a few seconds. By analysing these vapour trails you can work out what particle has just flown through your detector. It's like the particle physics equivalent of setting up a motion-activated camera and waiting for an animal to stroll past.

Anderson netted himself a Nobel Prize for using cloud chambers to discover the positron, but in 1936 something shot through his chamber without warning which nobody had ever seen. The shape of the vapour trail it left behind was characteristic of an electron, with one major difference – it was 207 times heavier.

This would be like setting up a motion camera in your back garden in the hope of observing a neighbourhood cat and accidentally capturing an image of a sabre-tooth tiger instead. Anderson had observed a fifth type of particle which nobody had predicted.

Rather disconcertingly this new particle, christened the muon, wasn't found in atoms, didn't contribute to radioactive decay and wasn't needed to explain any quantum phenomena. It had no purpose, yet it existed.

When he heard about the muon, Nobel Prize-winning physicist Isidor Rabi famously said, in annoyance, 'Who ordered that?',[40] because the muon changed the way we thought about particles. Apparently, the Universe was not the simple, elegant thing we had assumed it to be. There were particles which existed for no reason. They didn't do anything. They just were.

Shortly after, the American physicist Martin Perl discovered another pointless particle which was christened the tauon (or simply 'tau') which was three-and-a-half thousand times heavier than the electron.[41] And as other physicists joined the party with their own cloud chambers, it kept on happening. Soon there was the pion, the kaon, the lambda, the xi, the eta, until by the mid 1960s there were over four hundred particles identified.[42]

This was the physics version of an existential crisis. The Universe was apparently bubbling with stuff that has no function and we still haven't figured out what to do with that. We have managed to boil the particle zoo down to a handful of key players called 'the standard model' which contains a dozen fundamental particles which the others are composed from. But now we're stuck.

Some physicists are searching for a sleek theory which will account for all the different particles and why they have such apparently random masses. But other physicists see no reason to assume the Universe is neat or pretty at all. Perhaps the Universe is, like so much of science, simply a mess.

PART FOUR

Eurekas

··

'The man of talent is a marksman who hits a target others cannot hit – the man of genius is a marksman who hits a target others cannot see.'
Arthur Schopenhauer

'When the right idea finally clicks in place . . . one could kick oneself for not having the idea earlier.'

Francis Crick

'Eighty-eight miles per hour!!!'
Emmett 'Doc' Brown (*Back to the Future*)

THE NAKED TRUTH

I couldn't in good conscience write about Eureka moments in science without talking about the event which gave us the phrase itself. Although what was actually at stake in the anecdote is often misunderstood.

Sometime in the third century BCE the king of Syracuse, Hiero II, asked a goldsmith to make him a crown and supplied a lump of gold for the process. The goldsmith went away and returned the next month with a beautiful crown, but Hiero was then privately informed that the gold-smith had cheated him.[1]

Every metal has a unique density, a measure of how closely packed its atoms are. Gold's density is about 19 grams for every centimetre cubed. The density of silver, on the other hand is half that, so if you're working with a pot of molten gold you can easily scoop out a centimetre cubed and replace it with double that amount of cheaper silver. It will weigh the same, and by the time the metal sets it won't look any different. Who could possibly discern a single centimetre cube difference by eye?

A cunning goldsmith could use this technique to replace the king's gold, syphoning off some for himself, but how could you prove such a dastardly scheme? Hiero decided that to test the metal of the goldsmith's mettle he would have to turn to one of his relatives,[2] a man who had a reputation for scientific and engineering brilliance: Archimedes.

Archimedes is credited as having invented the screw pump – an early method of transporting water – the pulley system and most excitingly, a giant mechanical claw which was used to lift invading ships out of the city harbour.[3] (You'd need to be a time-travelling archaeologist to abso-lutely confirm the existence of that last one, but it's an exciting thought!)

Hiero had faith that Archimedes could devise a way of testing the crown's purity – the only stipulation was that Archimedes couldn't melt the crown down in order to work out its volume. It was a well-made piece of jewellery after all, so Archimedes would have to determine its composition without altering it.

Weighing the crown wouldn't do any good since it was identical to the gold the king had supplied, so the issue was finding out if it had the same volume. For a regular shape like a cube or a sphere there are reliable mathematical formulae to calculate volume, but for an irregular object like a crown there's nothing. How could Archimedes work out the true size of the crown?

The answer came to him in a moment of iconic simplicity as he climbed into his bathtub one evening and saw the water around him rise. An object submerged in fluid will displace the same volume as itself, so if he put the crown into water, he could measure how much the level rose and thus know the size. He was so excited by this realisation that he jumped out of the bath and went running through the streets of Syracuse butt-naked, shouting 'Eureka!' – 'I've found it!'

According to the historical record of this event written by Vitruvius, the crown displaced a larger volume of water than it should have, meaning that it had been puffed up with cheaper metal. The goldsmith's fate isn't recorded (although let's be honest, he was probably killed), but Archimedes got so excited that he went on to study fluid physics in detail, eventually leading to the discovery of Archimedes' principle.

Archimedes' principle is not the fact that an object submerged in fluid will displace an equal volume, as is often mis-stated. The Archimedes' principle says that when a fluid gets displaced the weight of it will be the same size as the buoyant force holding the object up. This realisation, combined with the idea of volumes displacing, is what determines whether something will sink or float.

Compare a boat and a pebble, for instance. The boat weighs a lot more, yet it will float while the pebble will sink. Archimedes' principle explains why.

Suppose the boat weighs a tonne. When we put it into water it applies a downward force on the surface and the water responds by pushing back with a force of one tonne. Now we apply the principle – if the force pushing up on the boat is one tonne, one tonne of water is going to rise.

A tonne of water has a volume of one metre cubed, so we know how much the water level will go up. If the boat has an overall volume *bigger* than a metre cubed the rising water will be smaller than the size of the boat, i.e. the boat will not be overtaken by water and it will stay afloat.

Now consider the pebble. Suppose it weighs a thousand times less than a tonne. When we put it into water the same thing happens: a thousandth of a tonne pushes up on the pebble, displacing a thousandth of a tonne of water. This amount is one litre and if the volume of the pebble is *smaller* than a litre, the water will overtake it, i.e. it will sink.

Putting it another way: what determines the floatiness or sinkiness of an object isn't weight. It's how compacted or spread out the weight is. Density is what matters, so if an object is denser than a fluid it will sink. If it's less dense, it will float. And speaking of great Eureka moments involving water . . .

SPLISH-SPLASH

Lonnie Johnson was born in Alabama in 1949 during the era of segrega-
tion. Being black, he went to an all-black school where he started learning about the great black inventor, George Washington Carver. A passion-
ate student of science, Johnson would conduct experiments in his home,
nearly burning his kitchen down after making rocket fuel in a saucepan,
and getting arrested for driving his home-built go-kart down the
highway.

When his school learned how gifted he was, they entered him as a
contestant in the state-wide Junior Engineering Technical Society Science
Fair, held at the University of Alabama (where Governor George Wallace
had once tried to prevent black students attending). Johnson was the
only black contestant and scooped first prize with an air-powered robot.
In your face, Governor Wallace.

Johnson eventually attended Tuskegee University, the home of his
hero George Washington Carver, on a scholarship and earned degrees in
both mechanical and nuclear engineering. His skill for physics and
invention quickly came to the attention of the US Air Force and in 1982
he was recruited to help in the development of the stealth bomber,
specifically to redesign the heat pump.

Heat pumps at the time used Freon gas™, but it was starting to
become apparent how damaging to the environment that could be. The
Air Force wanted Johnson to create a pump which could work with water
instead. This was no easy task because water has a completely different
viscosity from Freon, meaning the pump would have to be redesigned
from scratch.

One evening, Johnson brought his work home with him and hooked
up his pump design to the bathroom basin. When he turned the water

on, he turned the tap a little too forcefully and the water jetted out of the nozzle at such velocity that it hit the opposite wall. Johnson stared at the puddle and wondered: what if instead of a heat pump, he adapted his fluid nozzle to work as a water gun?

After tinkering with the design, Johnson filed a patent for his device and tried to find a distributor for his invention. For seven years he failed to get interest, until finally he got into a board room with the toy manufacturer Larami, who were immediately impressed. His water gun, originally made from parts designed to cool the US Air Force stealth jet, was marketed as the Super Soaker™, the best-selling water gun in history, with an estimated value of $1 billion today.

Johnson didn't stop there, though. After the enormous success of the Super Soaker™ he tweaked the design to fire foam darts and thus also invented Nerf™ toys as well, estimated to have a value of $460 million.

Johnson today has over two hundred patents to his name, and also helped design the fuel system for the Galileo space probe – as if being the inventor of Super Soakers and Nerf guns wasn't already cool enough.[4]

BOBBING FOR APPLES

The story of Archimedes running naked through the streets of Syracuse may not be entirely true, but the second most important Eureka moment in history *definitely* happened – and we know that because we have a first-hand account of it from the genius himself.[5]

This time we're on a farm in Lincolnshire where a young Isaac Newton is puzzled by something. The year is 1665 and England has been put under lockdown due to the Great Plague. Universities are shut and Newton has to make do with helping his brother, sisters and mother on their farm – which he was not cut out for.

The thing which was troubling him was that Archimedes' principle seemed to be missing something. Air is a fluid and objects move downward through it because they are more dense. That was straightforward.

In order to explain how an object would fall, you needed to know its weight as well as its size and the properties of the fluid it was moving through. But, thought Newton, the moon isn't falling towards the Earth, and surely that had to be denser than air.

As he pondered this, he happened to notice an apple falling from a nearby tree. As it did so, the answer dropped into place inside his head.

The version of the story in which the apple hit him on the head was an embellishment by the author Isaac D'Israeli, who claimed that, 'As he was reading under an apple-tree, one of the fruit fell, and struck him a smart blow on the head. When he observed the smallness of the apple, he was surprised at the force of the stroke. This led him to consider the accelerating motion of falling bodies; from whence he deduced the principle of gravity, and laid the foundation of his philosophy.'[6] That's not quite how it happened, but a falling apple *did* give Newton his breakthrough, as he related to his biographer William Stukeley.

An object will not speed up or slow down unless something is forcing it to. If you push an object, its natural inclination is to move in a straight line forever. On Earth, an object only comes to a rest because there are forces opposing its motion – usually air resistance and friction. But if you push something in an environment where no other force is acting (like space), it will keep going. This principle had been known in approximate form since the time of Galileo.

But when something falls it isn't just moving at a constant speed: it's getting faster. A simple way to prove this is to imagine dropping an apple from an inch off the ground and then dropping it from 20 feet. The second scenario will damage the apple more because the apple is slamming into the ground at a higher speed. That means the apple is speeding up as it falls and the longer it falls, the faster it goes. When things are falling, Newton realised, there is a force making them do so.

The reason objects fell wasn't *just* about the object and the fluid. The apple and the air were only part of the equation, and Archimedes' theory was incomplete. The bit that was missing, the bit everyone had overlooked for a couple of thousand years, was that there was a third object involved: the entire planet Earth.

Rather than being the unimportant final resting place of the apple, Newton realised that the planet played a crucial part in the event. The apple wasn't moving according to density alone – it was moving because the Earth was attracting it. The whole planet was exerting an invisible pull, forcing it to move in a line. Things fell because the Earth had a force field which pulled everything inward. He named this force from the Latin word for 'weight' – *gravitas* – gravity.

This explained why the moon didn't fall, even though it's denser than air. The moon and Earth both have an invisible gravity field around them and are pulling towards each other constantly. But they are in motion, so when the moon tries to fall towards Earth the latter moves out of the way, forcing the moon to fall in a new direction over and over until it ends up spinning around the Earth in a loop.

The Earth was doing the same thing with the sun: constantly trying to fall towards it as its trajectory gets forced into a circular fall. The sun, in turn, is falling in a circle around the centre of the galaxy . . . and so on to infinity.

Newton's discovery of gravity wasn't just about figuring out why things fell. It was the first example of an idea which connected the Earth to the heavens. The principles which apples obeyed were the same principles governing the movements of everything in the cosmos. Newton had come up with something truly astounding, a principle which everything in the Universe had to obey – a *law* of physics.

JUST BOBBIN THIS TIME

In 1846 the American engineer Elias Howe was trying to invent an efficient way to join two fabrics. He had been labouring on designs for years, but the solution finally came to him in a dream.

In this dream, the king of an unnamed country gave him twenty-four hours to invent a sewing machine or he would be executed. Howe failed to come up with an idea, just as in real life, so he was taken to the nearby killing grounds by the king's warriors.

As they marched him to his death Howe noticed that the warriors carried giant spears with holes in their tips. His eyes snapped open at four in the morning and he instantly began drawing. The idea was so clear all of a sudden – put a thread hole in the *tip* of the needle, not the other end (the way every other sewing needle in the world is made)!

If you put a thread through the needle's point, as it pushes through a layer of fabric it will automatically be dragged through to the other side. By having a rotating drum (called a bobbin) waiting with a second piece of thread to catch it, you can form a knot between the threads. When the needle retracts back, the knot will be secured, successfully forming a stitch. Howe had invented the sewing machine.[7]

IT'S ALL RELATIVE

The German-born physicist Albert Einstein was arguably the first international celebrity scientist. He was deservedly hailed as a legend in his own time, but one of the problems the press had with his ideas was describing them to the public.

While some are easy to explain others are extremely fiddly, like the discovery for which he was awarded the Nobel Prize – proving that light is made of particles called photons as well as being made of electromagnetic waves (for a brief a summary of how this paradox is resolved see Appendix 4 – or if you fancy giving me more money, buy my book about quantum physics: *Fundamental*).

His two grandest achievements, however, are his theories of relativity: special relativity (1905) and its expanded form general relativity (1916).

Special relativity deals with the way things change depending on how we are moving relative to each other, a topic which had fascinated Einstein since he was a teenager. Beams of light, made of particles called photons, always travel at 300 million metres per second. That's a robust finding with no exceptions. No matter what the beams of light have been ejected from, be it bulb or star, they always travel at the same speed. But why should that be the case?

The answer came to Einstein in Bern one evening in May 1905, as he made his way home from his job at the patent office. The town square has a large clock tower at its centre and as it struck the hour, Einstein stopped in his tracks. Sadly the exact time isn't recorded, but as he looked up at the clock, he had his Eureka moment.[8]

Imagine travelling away from the clock and looking at it over your shoulder. Photons coming from the clock would land on the retina of your eye, telling you the time. But suppose you were moving away from the clock at 300 million metres per second. What would happen?

If you're travelling at the same speed as photons of light, they would not be able to catch up with your eyes. So even though the face of the clock would be constantly updating, you wouldn't be able to see it happen. You'd be moving faster than the information could reach you. Even an hour later, you would see the same time frozen on the clock face.

It wouldn't just be the clock that would be frozen, either. Everything else taking place behind you would appear locked in a tableau because no new photons would be reaching your eyes. From the perspective of a photon, Einstein realised, the outside Universe appears still. Photons do not experience time.

Now extend this idea and imagine a different scenario – that you're travelling away from the clock at *almost* 300 million metres per second. You'd see the clock hands ticking this time, but the photons would be reaching you with a delay in between each tick – the clock would appear to move, but very slowly. In other words, if you stand next to a clock it will pass at a normal rate but if you move away from it faster and faster it will slow down from your perspective.

Time, Einstein realised, was not a constant that everyone in the Universe experiences identically. It is dependent on the speed you're moving at relative to something else. The faster you move compared to anything else, the slower that thing would appear to be.

The clock tower would still be ticking at one second per second but as you sped up, its time relative to you would be slowed down. You'd feel normal while the rest of the world would look wrong, but from the clock's perspective time would be passing too quickly for you.

Two people holding clocks and moving in different directions at different speeds would disagree about how much time had passed. Time itself was therefore relative.

The theory which emerged from this realisation (special relativity) is a mathematically fiendish exploration of light, time and movement which finally explained *why* light always moves at 300 million metres per second. The faster an object moves, the slower it experiences time.

Photons are, in a sense, trying to go as fast as possible. It's just that when they hit 300 million metres per second, time stops passing for them so they can't go any faster. So this isn't really 'the speed of light': it's the universal speed limit itself. Nothing can travel faster than 300 million metres per second because nothing can move through time slower than zero.

Einstein eventually expanded special relativity into a beefier theory which included gravity, mass and energy, but that was the result of hard work and mathematical agony. Special relativity, where it all began, was the result of simply looking up to check the time.

ANOTHER LIGHT STORY

Einstein and the Bern city-centre clock wasn't the only time a beam of light from an object gave someone an invaluable Eureka moment. In December 1934 the road contractor Percy Shaw was driving home from The Old Dolphin pub in Bradford, Yorkshire. It was late, foggy and there were no street lights due to maintenance on the roads. Shaw was finding it impossible to see where he was going, when suddenly two bright lights shone out at him from the darkness ahead.

A tabby cat was sitting on a brick wall staring at his approaching vehicle, and the headlights of Shaw's car had caught the retinas of the cat's eyes. If it hadn't been for that reflection, Shaw would not have seen the wall and would have suffered a potentially fatal crash.

If only, he thought, he could persuade a small army of cats to line up along the middle of the road and guide him home. Then (presumably remembering that his job was to build roads) he had an idea.[9]

Shaw's cat's eye design uses small reflective surfaces surrounded by rubber to catch beams of approaching car headlights. Cleverly, the cat's eye also has a chamber which fills with water during rain and when a car drives over it, the water squirts onto the reflective surface, cleaning it in the process.

During the Second World War cat's eyes became enormously useful because we didn't want to have street lights on, showing the outlines of our cities. But we could still have cars driving around at night clearly seeing where the road was, thanks to reflections in the cat's eye devices which Nazi planes above couldn't see.[10]

A tremendous invention, owing its existence to a very startled cat. Had the cat been facing the other way of course, Shaw would no doubt have invented the pencil sharpener.[11]

The scandal of life

During the early 1950s a race was taking place between two of the top biochemistry labs in the world. The goal was to be the first to work out the structure of a chemical called deoxyribose nucleic acid (DNA).

In 1944 an experiment conducted by the Canadian-American physician Oswald Avery had shown that DNA was almost certainly the genetic molecule; the chemical responsible for containing hereditary information.[12] The problem was that nobody could figure out how this was possible.

DNA was known to contain a few simple building blocks, but not much else. It contained four molecules called chemical bases: adenine, thymine, guanine and cytosine (ATGC for short) and a long thread-like component made of phosphate and ribose (ribose usually has a lot of oxygen in it, but these ribose chains had fewer, hence 'deoxygenated ribose'). But that was it. It was too simple. A deoxyribose/phosphate thread and four chemical bases? How was it possible to code the entire library of life with so crude an alphabet?

At King's College London the research was being led by Rosalind Franklin and Maurice Wilkins, while at Cambridge, rival work was being conducted by Jim Watson and Francis Crick. The Cambridge team of Watson and Crick got on brilliantly, but the London team of Franklin and Wilkins did not. There have been numerous accusations of misogyny and bullying levelled at Maurice Wilkins over the years[13] but he always denied them, writing off the friction with Rosalind Franklin as a simple personality clash.[14]

It was Franklin who first suggested DNA was a spiral shape with a deoxyribose/phosphate thread forming a helix, with the bases A, T, G and C sprouting from it like clothes pegs on a washing line.[15] Watson then 'had the same idea' (immediately after hearing Franklin give a lecture about it) so he and Crick set about building 3D models to find out which parts of DNA could sit next to each other.

Watson and Crick initially announced they'd cracked the structure and invited Wilkins and Franklin up to Cambridge to have a look. Not only did Franklin point out that parts of their model were off by a factor of 90 per cent, she also explained that their model was inside out. Watson and Crick had put deoxyribose/phosphate backbones on the *inside*, corkscrewing together, which has the irritating property of being impossible. Deoxyribose/phosphate chains repel each other, so Watson and Crick's DNA would have exploded.

Franklin spent the next three years carefully gathering all the information she could about DNA, calculating angles between parts of the chain and eventually managing to capture a silhouetted image of it in side profile. During those years, Watson and Crick were having no luck working out the structure (they had been banned from working on it after making fools of themselves), but eventually Watson hit on the brilliant idea of travelling down to London to talk to the King's College team and see what they'd come up with.

After getting into a heated row with Franklin in her lab, during which Watson accused her of being unable to interpret her own data, Wilkins tried to console him by showing him Franklin's silhouetted image of DNA. Watson sketched out a copy and took it back to his headquarters in Cambridge. Watson and Crick also got their hands on Franklin's numerical data because their supervisor worked for the organisation she had submitted her unpublished work to for verification.

Now, to be clear, in science there's no such thing as plagiarism. Science is about facts and facts aren't property. If anything, you want other people to use your ideas because that's how you test them, so Watson and Crick weren't doing anything illegal by getting a secret backdoor key to Franklin's research. You don't go to science prison for using someone's data without asking them . . . but it's just . . . not . . . honourable.

Watson and Crick barrelled ahead with their sneakily obtained data, but still could not solve it. At all. DNA was apparently a double helix with two backbones of deoxyribose/phosphate winding round each other on the outside, while the bases were sticking out at right angles on the inside. It was like a twisted ladder with the deoxyribose/phosphate forming the rails and the bases forming the rungs.

Crick had deduced that the deoxyribose/phosphate rails were arranged in what's called antiparallel configuration (see Appendix 5), but the bases were all different sizes. There was no way of getting them to stack up inside a double helix. It would be like trying to line up a bunch of pennies inside a tube except all the pennies are different diameters so you can't get a smooth cylinder. How could you fit four differently sized bases inside without it being buckled?

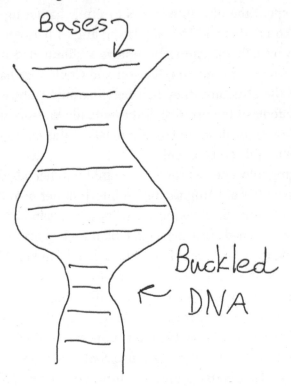

Then came the Eureka moment, which is a polite way of saying it. A better way of saying it might be: someone else from their department happened to walk through Watson's office and told him the answer. It's so ludicrous that if you were to write it as a film script, critics would complain that the finale was too much of a coincidence. They'd throw popcorn at the screen and moan that the writers were lazy – except that's the way it happened.

One evening, Watson was pulling his hair out looking at his models of the bases, when another scientist, Jerry Donohue, happened to strike up a conversation with him. When Watson outlined the problem he was having, Donohue just shook his head in amusement.

Donohue happened to be the world's expert on the base chemicals Watson and Crick were studying. Not only that, he knew the explanation for why they were getting nowhere – the textbooks had it all wrong. The four bases were always described as coming in a variety of shapes but actually they only came in two, one big and one small.[16]

After dropping this casual bombshell, Donohue walked out and Watson stared down at his models, making the final deduction. A and T came in one size while G and C came in another. When you lined up the bases long to short they always had the same width across.

The bases were paired up in a constant width running the length of the DNA. The ordering inside the spiral could be complicated, storing vast amounts of information, but from the outside it would look simple and uniform.

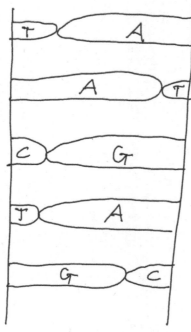

Watson and Crick had beaten Franklin and Wilkins to the punch. The two teams published their findings together in the same journal as a supposed show of scientific solidarity, but Watson and Crick undermined the spirit of this by declaring in their paper that they 'were not aware' of Franklin's data.[17] There isn't really any good scientific reason to put something like this in your paper unless you're really wanting to make sure people think you came up with the idea on your own.

Watson, Crick and Wilkins eventually won the Nobel Prize for elucidating the structure of DNA. Rosalind Franklin had sadly died from ovarian cancer four years before and did not get the recognition she patently deserved.

In later years, Wilkins and Watson's behaviour and mistreatment of Franklin has come under heavy fire from the scientific community, and Watson was coincidentally stripped of many titles and positions for making derogatory comments about women,[18] Jewish people[19] and black people.[20] However, he did get the critical Eureka which solved DNA . . . after Jerry Donohue unexpectedly told him what it was.

TOILET CLEANER

All right, it might not be as earth shattering as the law of gravity or the secret of life, but toilet cleaner is a wonderful discovery which shouldn't be disregarded. Lots of people have toilets and lots of people want to keep them clean, so let's not ignore the man who invented toilet cleaner – and the Eureka moment which inspired him.

After the First World War, Harry Pickup was given the responsibility for cleaning out London factories where the military had been making explosives. While emptying one such factory, Pickup came across a powdery bomb residue which he tried to spritz away with water. In the process of dissolving it, he noticed it was stripping limescale off the surrounding surfaces.

Limescale is a build-up of calcium carbonate, completely harmless but unsightly and difficult to remove. The powdery substance Pickup was trying to clean was sodium sulfate, a by-product formed in the production of nitric acid (which is itself used for bomb production). Sodium sulfate is water soluble and when dissolved, it will convert to sulfuric acid, which Pickup discovered would dissolve limescale.

The convenient thing about Pickup's discovery is that nobody wants to be lugging around bottles of highly corrosive sulfuric acid whenever they want to clean their toilet. But if you put powdered sodium sulfate into the water directly, it converts into sulfuric acid below the surface, saving you from having to pour it yourself. This innocent observation of seeing the limescale vanish gave Pickup the idea for a cheap and effective toilet cleaner.

He went into business, marketing his limescale-removing powder as Harpic™ (from his name HARry PICkup).[21] As I say, perhaps not an Einstein-level moment, but you have to respect a man who thinks to himself: 'I wonder what happens if I put bomb chemicals down a toilet?'

Expecting the unexpected

....

In the first draft of the book, I originally planned to open with a quotation attributed to my favourite science writer, Isaac Asimov:

> *'The most exciting phrase to hear in science, the one that heralds new discoveries is not "Eureka!" but "That's funny . . ."'*

It's a great quotation, and although it's exactly the kind of thing Asimov would have come up with, nobody really knows who said it. Fleming reportedly remarked, 'That's funny,' when he discovered penicillin, but the rest of the sentence is an embellishment.

In the end, I went with a quotation from G. K. Chesterton because it felt more in keeping with the tone of the book, not to mention the fact that I wanted to include a few Eureka stories after all. But I'd feel bad if I didn't pay at least some kind of homage to Professor Asimov, so I decided to finish things off with another idea he proposed.

In his seminal Foundation novels Asimov imagined a future in which humanity has found a way to predict its own fate. While it's impossible to foresee the actions of an individual, a population's overall trends might be more manageable – in the same way that it's difficult to predict the skittering of a single atom, but the flowing of trillions of them in a river is straightforward.

Asimov called this fictitious discipline of future prediction 'psycho-history', and it's used by a group of scientists to guide humanity through bottlenecks that threaten our extinction. A spanner is thrown into the works in the second novel, however, when their predictions start deviating from their trajectory, due to the machinations of a mysterious figure who can control people's minds and influence their actions.

This powerful schemer, known as 'The Mule', starts vying for political control, using his telepathic influence to manipulate Government dignitaries and leaders. Soon, the carefully calibrated equations of psychohistory are under threat and the scientists have to find a way of correcting course before The Mule threatens the stability of the galaxy.

I've often felt that The Mule was Asimov's metaphorical way of representing moments in history which nobody saw coming but which ended up changing everything. A butterfly flaps its wings in Brazil, as the saying goes, and we get a tornado in Texas.

In the real world a discipline similar to Asimov's psychohistory actually exists, although it's still in its infancy. The field of 'cliodynamics' is an attempt to combine economics, sociology, anthropology and evolutionary history to mathematically model the ebbs and flows of our human endeavour. But, as with Asimov's novels, cliodynamics faces an insurmountable problem: sometimes random stuff happens.

As we've seen over and over throughout this book, however, whenever our plans go haywire the overall flight path of humanity has still been towards enlightenment. Humans have a remarkable ability to learn from our mistakes and to incorporate surprises into what we know. Despite our primitive beginnings we've moved closer to a world where things are better – and we will continue to do so. Is the world perfect? Absolutely not. But have we come a long way? Hell yes. And we'll never stop.

We'll keep our eyes out for discoveries and we'll keep paying attention to baffling readings. We'll keep getting knocked back by malady and misfortune, but we'll keep dusting ourselves off and learning from the failures.

It's impossible to predict how we will overcome the many challenges headed our way, because a lot of the solutions will probably not be deliberate. But whether it means to or not, science will save our species.

The periodic table of 'huh?'

There are ninety-two naturally occurring elements in the periodic table, and sixty-eight of them were discovered by chance (two of them independently involving seaweed). Even the periodic table itself was hit upon in a curious way while Dmitri Mendeleev was trying to figure out how the elements should be arranged.

He invented a game of playing cards with element properties written on them, which he would try to arrange in patterns (the nerdiest game of solitaire ever invented). After pulling a seventy-two-hour-no-sleep study bender, Mendeleev collapsed and had a dream in which his playing cards arranged themselves into a pattern – the periodic table.

Mendeleev's table was a game changer for chemistry, not just because it allowed us to group elements by chemical property but because it left gaps for elements we hadn't yet discovered.

The story of how chemical thinking evolved deserves its own book (see *Elemental* – and buy copies for your friends), but since so much of it happened by accident it definitely deserved to be addressed in this one. However, writing a whole chapter which consisted of ninety-two short vignettes seemed out of kilter with the other sections, so my editor suggested sticking it here after the Epilogue but before the Appendices. I think she was right.

Rather than go into detail for every element, I'm going instead to provide a whistle-stop tour of each one and how they were isolated. I've categorised them into seven types.

Total accident Nobody was looking for the element or even suspected it was there, it just showed up.

Impurity The element was an impurity mixed in with something the experimenter was supposed to be studying.

Breakdown surprise Someone was breaking something down to see what it was made of (through various methods, e.g. acid, burning, electricity) and found something unexpected.

Spectroscopy surprise The element was hidden inside something else and emitted an unexplained beam of light when heated.

Blind reaction Someone was randomly reacting stuff together and got lucky.

Predicted The element was either predicted by Mendeleev's table or strongly suspected to exist.

Known to the Ancients The element is recorded in ancient records so we don't know how it was identified.

1 Hydrogen (H) Blind reaction

In 1671 British scientist Robert Boyle was experimenting with hydrochloric acid and tipped some iron filings into it. When he did so, it 'belch'd up stinking fumes'[1] that he considered so unpleasant he didn't want to write any more about them.

It's not clear what this awful-smelling gas was (my guess would be that sulfur impurities, often found in iron, reacted with the acid to produce hydrogen sulfide, the smell of rotting eggs), but the other gas he made ended up coming into contact with a candle, causing a violent explosion.

A rigorous experiment on this gas was eventually carried out by the English philosopher Henry Cavendish in 1766 in which he collected bottles of it and found it would ignite in air to create water.[2] He named it from the Greek words *hydro-genes* meaning 'water-maker'.

2 Helium (He) Spectroscopy surprise

In 1868 the French astronomer Pierre Janssen was studying light coming off the sun during an eclipse. Most of the light he observed matched light produced by hydrogen on Earth, but there was one unfamiliar beam with a wavelength of 587.49 nanometres which didn't match anything else on record.

Initially mistaking it for sodium, he consulted with his friend, the English astronomer Norman Lockyer, who concluded that he had discovered an extraterrestrial element, which they named after Helios, Greek god of the sun. Helium is thus the only element to have been discovered in space before it was found on Earth.[3]

3 Lithium (Li) Total accident

In 1800 the Brazilian statesman José Bonifácio de Andrada e Silva was visiting a mine on the island of Uto when he came across a rock that had not been classified. He named it petalite from the Greek *petalon*, meaning 'leaf-like', due to the way the rock would split when struck.

At some point during his trip Silva dropped a piece of petalite into a fire, where it turned the flames red; but nobody believed him, or even believed petalite existed. It wasn't until seventeen years later that the Swedish mineralogist E. T. Svedenstjerna rediscovered petalite and its red flames. Svedenstjerna sent a sample to his friend, chemist Johann Arfvedson, for analysis, and Arfvedson identified the culprit – the element lithium, which he named from the Greek *lithos* meaning 'stone'.[4]

4 Beryllium (Be) Predicted

In 1798 the French mineralogist René Haüy had come to the conclusion that emerald and beryl were probably not the same gemstone. They both contained aluminium, oxygen and silicon, but he was convinced there was something else. In order to prove he was right he sent samples to his friend Louis Vauquelin, and asked him to break them down. Vauquelin found that beryl did indeed contain a new element. Originally he proposed the name glucinium from the Greek for 'sweet' (*glykós*) because it formed compounds with a sweet taste, but beryllium was settled on after the gems that had contained it (from the Greek *beryllos* meaning 'turquoise gemstone').[5]

5 Boron (B) Breakdown surprise

In 1808 the English chemist Humphry Davy was testing the electrical conductivity of solutions. He knew that if you stuck two ends of a battery into most liquids the current would transfer easily through the fluid. But when he tried it with solutions of borax (a mineral that had been known for at least two thousand years, so the etymology is lost), he found a brown residue building up at one end of the battery.

The reason this happened was that there were charged particles of an unknown element floating around in the solution. When he passed a current through it, these particles were drawn to one end of the battery, forming pure clumps of the element boron.[6]

6 Carbon (C) Known to the Ancients

Although – since this is the main element left over when any biological material burns – there's a good chance carbon was discovered when someone or something got caught in a fire/a lightning strike and left behind some charred black remains. I'm counting it as a probable accident.

7 Nitrogen (N) Total accident

In 1772, the Scottish chemist Daniel Rutherford's pet mouse died. It died because he had put it inside an airtight box and the poor thing ran out of oxygen, converting it to carbon dioxide. Rutherford tried burning a candle inside the carbon dioxide and found it didn't last very long because there was no oxygen to react with the molten wax.

He decided to take the gas from inside the box and pass it through limewater, a substance which absorbs carbon dioxide, but found there was still a gas present even after the CO_2 had been removed. There was some other gas present in air that the mouse hadn't absorbed.[7] He called it 'noxious air' but it was eventually renamed by the French chemist Jean Chaptal when he learned it could be used to generate *nitre*, the French for saltpetre, hence *nitre-genus* . . . nitrogen.[8]

8 Oxygen (O) Various

Several scientists can claim to be the first to isolate oxygen. In 1604 the Polish alchemist Michael Sendivogius discovered that burned saltpetre produced a gas (which we now suspect was oxygen).[9] Joseph Priestley, an English chemist and theologian, discovered in 1774 that mint leaves breathe out a gas which could keep a mouse alive – he also killed several of them in the same manner as Daniel Rutherford.[10]

In the late eighteenth century the French chemist Antoine Lavoisier discovered that metals got heavier when burned in air, showing they were combining with an invisible element (an experiment involving a 9-foot magnifying glass used to focus the sun's rays to ignite a plate of mercury).[11] It was Lavoisier's name for the element which stuck, from the Greek *oxys-genus* meaning 'acid maker', since many of the acids he knew about contained oxygen.

9 Fluorine (F) Predicted

Hydrofluoric acid, derived from dissolving fluorite minerals in water, has been used for glass etching since 1670 when Heinrich Schwanhard, a German glass cutter, accidentally damaged his spectacles with it. Many chemists tried to isolate the ingredients of this highly corrosive acid (some dying in the process) but it was finally achieved in 1886 by French chemist Henri Moissan.[12] It gets its name from the Latin *fluere* meaning 'flow', since fluorite minerals were used to make smelting mixes flow better.

10 Neon (Ne) Total accident

In 1894 the English chemist William Ramsay attended a lecture given by Lord Rayleigh in which Rayleigh described an anomaly he had found. Nitrogen gas obtained from chemical reactions had a different density from nitrogen gas extracted from the air. Air was known to contain oxygen and nitrogen, but Rayleigh couldn't account for why atmospheric nitrogen had different properties from chemically generated nitrogen.

Ramsay decided to cool down some air into liquid form and slowly separate the oxygen and nitrogen out to study them closer. He discovered that there was still a puddle of liquid left in the tube afterwards – something which was clearly present as a trace amount.

Further inspection showed that he was wrong. There wasn't a trace element in the air – there were *five* trace elements. These stable gases had gone undetected in previous experiments because they were completely unreactive so had always been overlooked.

Discovered simultaneously in Ramsay's puddle, they were christened noble gases since they didn't interact with the 'common' elements. There was helium (already discovered on the sun), neon, argon, krypton and xenon. Neon got its name from the Greek *néos* meaning 'new'.[13]

11 *Sodium (Na)* Breakdown surprise

Isolated by Humphry Davy in 1807 using a similar approach to the one he used for boron – only this time he was testing the conductivity of lye solution. Quicklime was often consumed medicinally as it was believed to treat headaches, hence its name which comes from the Latin *sodanum* meaning 'headache medicine'.[14]

12 *Magnesium (Mg)* Breakdown surprise

Once again, Humphry Davy with his battery, this time in 1808 when passing electrical current through a solution of Epsom salts from the town of Magnesia in Greece.[15]

13 *Aluminium (Al)* Predicted

This time Humphry Davy predicted the element in 1808 but couldn't extract it successfully. It was finally isolated in 1824 by the Danish chemist Hans Ørsted (who also discovered the electromagnetic effect mentioned in Part Three) by reacting aluminium chloride with more reactive metals to kick out chlorine and leave the aluminium on its own.

Its name comes from the Latin *alum* meaning 'bitter', due to the taste of its compounds.[16]

14 Silicon (Si) Predicted

Swedish chemist Jöns Jacob Berzelius was searching for this one because there was a gap in Mendeleev's periodic table between elements 13 and 15.* He found it in 1823 using a piece of flint, for which the Latin name is *silicis*.[17]

15 Phosphorus (P) Total accident

The very first element to be isolated in modern times was found by German alchemist Hennig Brandt in 1669. Brandt was trying to obtain gold and had settled on the idea of boiling his own urine to see if it contained any. Urine has a golden colour, so Brandt decided to boil away the water in the hope of extracting precious metal.

Instead of gold, Brandt obtained a yellow powder which, for some reason, he roasted with pure charcoal. This burned away the urea in the powder, breaking it down and exposing the other substance present – phosphorus.

This new whiteish-blue powder glowed in the dark (hence *phosphorus* from the Greek 'bringer of light') and would catch fire easily, even burning under water. Brandt started producing phosphorus on an industrial scale by importing barrels of urine from the military and boiling it in his home lab, spending his wife's fortune in the process.[18]

---

* Mendeleev's table actually arranged atoms by mass not atomic number (counting in protons wasn't possible yet), but usually elements next to each other in atomic number are next to each other in mass too. Elements 13 and 15 have masses of 27 and 31 respectively, so it was a fair bet there would be something in between. Silicon has a mass of 28 so slotted in perfectly.

16 *Sulfur (S)* Known to the Ancients

17 *Chlorine (Cl)* Blind reaction

Found in 1774 by the Swiss-German pharmaceutical chemist Carl Scheele who reacted half an ounce of manganese oxide with hydrochloric acid before leaving the bottle open in a warm place. He noted the smell of the gas coming off the reaction, and although he didn't realise it was an element (Humphry Davy made that link) I think he deserves the credit for isolating it first.[19] Chlorine is faintly green, from which it gets its name (Greek *chloros*, meaning 'green').

18 *Argon (Ar)* Total accident

Discovered as part of William Ramsay's puddle. Named from *argos*, the Greek for 'lazy', since it is unreactive.[20]

19 *Potassium (K)* Breakdown surprise

Once again Humphry Davy and his battery in 1807. This time electrocuting a solution of potash, which gets its name from the fact that it is literally the ash found in a pot after you burn wood in it.[21]

20 *Calcium (Ca)* Breakdown surprise

Humphry Davy with his battery, this time trying it on limestone (calcium carbonate). Gets its name from the Latin *calx*, meaning 'limestone'.[22]

21 *Scandium (Sc)* Predicted/Breakdown surprise

Dmitri Mendeleev predicted this element's existence because there was a gap between 20 and 22. However, the actual discovery was made in 1879 by the Swedish chemist Lars Nilson who hadn't heard of the periodic table (it had only been invented ten years previously). He was breaking down the mineral ytterbia to find its components and inside found the element which he named after Scandinavia.[23]

22 *Titanium (Ti)* Blind reaction

Isolated in 1791 by the English priest William Gregor from black sand he was reacting with acid. He named it menachanite after the Manaccan valley in Cornwall where he discovered it, but the eventual name settled on was proposed by German chemist Martin Klaproth and based on the Titans, sons of the Earth in Greek mythology.[24]

23 *Vanadium (V)* Breakdown surprise

Isolated in 1801 by the Mexican chemist Andrés Manuel del Rio while analysing the red mineral vanadinite (named from the Norse goddess of beauty Vanadis due to its beautiful appearance). His discovery was initially dismissed as incorrect and he was told he'd misidentified a sample of chromium until further inspection thirty years later vindicated him.[25]

24 *Chromium (Cr)* Breakdown surprise

Isolated in 1794 by Louis Vauquelin by breaking down the mineral chromite, commonly used in red paints, which he had obtained from a Siberian gold mine. It gets its name from *chroma*, the Greek for 'colour', due to its numerous colourful compounds.[26]

25 *Manganese (Mn)* Blind reaction

Discovered in 1774 by Johan Gahn when heating samples of the mineral manganesia (from the Greek town of Magnesia).[27] Nobody is sure how the word ended up with two 'n's in it.

26 Iron (Fe) Known to the Ancients

27 Cobalt (Co) Predicted . . . ish

The second element to be isolated after phosphorus, this time by the Swedish chemist George Brandt (no relation to Hennig, just a great coincidence) in 1735. Brandt was deliberately looking for it because samples of metallic ore from German mines were thought to be haunted by goblins (*kobolds*).

Kobold minerals were used to dye glass blue, showed spooky magnetic properties, didn't smelt normally and people who worked with them often got inexplicably sick (a result of cobalt poisoning). Brandt was trying to rule out goblins and found the metal responsible instead.[28]

28 Nickel (Ni) Breakdown surprise

Discovered by the Swedish mineralogist Baron Axel Cronstedt in 1751 when he was trying to extract copper from some stubborn ores and found a strange metal that had properties *similar* to copper but not exactly the same. He called it 'copper of the devil' which in German is *kupfer*-nickel.[29]

29 Copper (Cu) Known to the Ancients

30 Zinc (Zn) Known to the Ancients

31 Gallium (Ga) Spectroscopy surprise

French chemist Paul-Émile Lecoq was analysing the light spectrum of a mineral called pierrefite when he observed two violet beams that didn't match anything previously documented. After boiling down 52kg of the stuff in lye he was able to purify it into a new element.[30,31] He named it gallium from *Gallia*, the Latin for France, although some suspected he was being sneaky and naming it after himself. His surname is French for 'cockerel', and in Latin that translates as *gallus*.[32]

32 Germanium (Ge) Total accident/Impurity

In September 1885 a mineshaft collapse occurred in the Himmelsfürst mine in Saxony, Germany, half a kilometre underground, yielding seams of a new silverish ore called argyrodite. In February the following year, Clemens Winkler was trying to extract silver from this ore and found a pesky impurity mixed in. He named it germanium after his homeland, Germany (*Germania* in Latin).[33]

33 Arsenic (As) Known to the Ancients

34 Selenium (Se) Impurity

In 1817 the Swedish chemist Jöns Jacob Berzelius and his friend Johan Gahn were running a sulfuric acid plant near Gripsholm. The method they were using was, and still is, the standard way to make sulfuric acid. The mineral pyrite (chock-full of sulfur) is roasted in air until it forms dark red clouds of sulfur dioxide. Then pump in a vapour of cold water which absorbs the sulfur dioxide, forming sulfuric acid in the process.

But when their plant started up, they found a mysterious red powder clogging up the reaction chamber. The pyrite they had been using turned out to have an impurity in it, which was left behind when the sulfur was burned away.

Berzelius examined the red powder and found that although it was similar to known elements such as tellurium, when it was burned it smelled powerfully of horseradish – this wasn't a known material at all. He settled on the name selenium from Selene, Greek goddess of the moon, seemingly because he liked the way it sounded.[34]

35 Bromine (Br) Breakdown surprise

Discovered by the French chemist Antoine Balard in 1825 while burning fucus seaweed. Fucus was a known source of iodine, an important dietary supplement, but when Balard burned the seaweed to ash, dissolved it in water and then rinsed it with starch, he discovered a tiny amount of brown liquid with a 'penetrating smell'. He named it from *bromos* the Greek for 'stench' on the advice of his friend M. Anglada.[35]

36 Krypton (Kr) Total accident

Discovered as part of William Ramsay's puddle and gets its name from *kryptos*, the Greek for 'hidden'.[36] Many years later, comic writers Jerry Siegel and Joe Shuster picked it as the name for Superman's homeworld because it sounded unusual and science-y.[37]

37 Rubidium (Rb) Spectroscopy surprise

Discovered in 1861 by German chemist Robert Bunsen (he of the burner and also inventor of the spectroscope). While heating a piece of lepidolite, he observed a deep red light emitting from the rock. It gets its name from the Latin *rubidus* meaning 'dark red'.[38]

38 Strontium (Sr) Breakdown surprise

Humphry Davy and his battery again in 1808. This time with a sample of ore from the village of Strontian in Scotland.[39]

39 Yttrium (Y) Breakdown surprise

In 1789 the Swedish lieutenant Carl Arrhenius was touring a mine in Ytterby where he came across a black rock he hadn't seen before (in fact, it was the same mine where Scandium would later be discovered). Arrhenius sent the black rock to his scientist friend Johan Gadolin who confirmed the presence of a new element which he named after the mine. Although it was so hard to purify, it wasn't actually isolated until 1828 by Friedrich Wöhler, a German chemist.[40]

40 Zirconium (Zr) Breakdown surprise

Similar story to yttrium. The German chemist Martin Klaproth had successfully identified a new element hiding inside a piece of the brilliantly named mineral, jargoon, in 1789 but couldn't isolate it. In 1824 Jöns Jacob Berzelius managed to force it out of hiding. It gets its name from the Persian *zargun* which means 'golden' since it has a vaguely gold appearance.[41]

41 Niobium (Nb) Breakdown surprise

Isolated by the English mineralogist Charles Hatchett in 1801 from a sample of the mineral columbite. Its chemical properties are similar to the element directly below it on the periodic table, tantalum, and since tantalum gets its name from the Greek god Tantalus, this new element was named after Tantalus's daughter, Niobe.[42]

42 Molybdenum (Mo) Breakdown surprise

The mineral molybdena was originally believed to contain lead. Its name comes from the Greek for lead-like, *molybdos,* and there was a good reason for the confusion. Molybdena is actually a blend of molybdenum sulfide and graphite. Graphite looks identical to lead, so when people were previously extracting what they thought was lead they were actually pulling out graphite and discarding the molybdenum sulfide.

In 1776 Carl Scheele was sent a piece of molybdena and told to extract the lead but Scheele was a good chemist and knew exactly how to get lead out of minerals. When he failed to do so with molybdena he realised everyone was mistaken – it didn't contain lead at all, which got people intrigued about what it really was. The element itself was finally isolated in 1781 by the Swedish chemist Peter Hjelm.[43]

43 Technetium (Tc) Predicted/Total accident . . . twice

This was an element predicted to be sitting in the gap between the known elements 42 and 44 on Mendeleev's periodic table, but it proved the most elusive. In fact it was the last 'gap' on the table to be filled.

The reason is that technetium is unstable and doesn't last for very long. The reason for *that* is a mystery. We know technetium atoms form with a nucleus overloaded with neutrons – but if you ask a physicist what makes technetium unstable they'll wave their hands and talk about 'nuclear stability' which doesn't really answer the question as to *why* that specific number of protons and neutrons is unfavourable. It just is, for some reason.

Technetium was first manufactured in 1934 by Fermi and his team (as we saw in Part Three), but they mistook it for another element, leading to the false Nobel Prize confusion.

The weird bit is that it was then manufactured and discovered by accident again in 1937. This time the American physicist Ernest Lawrence was working with a particle accelerator and getting rid of old molybdenum metal from inside the core because it had become radioactive. His long-distance friend Emilio Segrè asked if he could have a piece of it for analysis and Lawrence shruggingly agreed, posting him a sheet of radioactive molybdenum in the mail all the way from California to Italy.

There, Segrè inspected the molybdenum and found that some of its atoms had radioactively mutated into technetium, finally confirming its discovery. Lawrence hadn't realised he'd made technetium by mistake and sent a sample of it across the world via airmail. It was named after the Greek *teknetos* meaning 'artificial'.[44]

44 Ruthenium (Ru) Breakdown surprise

In 1844 Tsar Alexander I insisted that platinum from the Ural mines in Russia be used in the manufacture of the rouble coin, which meant a lot of ore from the mines was stripped for its platinum, discarding all the other chemicals inside. Russian coin factories therefore had an enormous build-up of useless mineral by-products, which chemist Karl Klaus decided to analyse, leading him to discover ruthenium.

Excitedly, he sent a sample of it to Jöns Jacob Berzelius for confirmation but became too impatient to wait for the reply and published his results anyway. Berzelius was not impressed by the sample and publicly dismissed it as 'dirty salt'. Fortunately for Klaus, however, other chemists came to his rescue and confirmed its identity.

He named it after the Latin for Russia, *Ruthenia*. (Incidentally, the chemist Gottfried Osann had previously isolated ruthenium in 1827 but had been unable to prove it since such a small amount had been obtained. By coincidence he *also* named it ruthenium, only this time it was after the mountains where the ore was mined.[45])

45 Rhodium (Rh) Impurity

An 1803 discovery by the English mineralogist William Wollaston who was trying to extract platinum from ores to sell on. He came across a black impurity in the platinum which turned faintly pink in water.[46] He named it for the Greek *rhodon* meaning 'rose' (not – as I misheard when I first learned about it – after Rodan the pterodactyl monster who fights Godzilla).

46 Palladium (Pd) Breakdown surprise

Discovered in 1802 by William Wollaston from the same platinum ore that led him to rhodium. He named it after Pallas, the recently discovered asteroid, although for some reason known only to him, he pretended not to be its discoverer. Instead, he sold ribbons of it through a shop in London, refusing to name the person who had given it to him – simply explaining that it was an element called 'new silver'.

When the chemist Richard Chevenix publicly refused to acknowledge palladium as an element, believing it to be an alloy of platinum and mercury, Wollaston anonymously offered a £20 reward (£500 in today's money) to anyone who could manufacture it. If it was just a platinum/mercury alloy it should be easily duplicated. When nobody was able to do so, the Royal Society agreed palladium was an element rather than an alloy, at which point Wollaston revealed himself as the masked chemist.[47]

Why Wollaston went on this circuitous route is unclear. Perhaps it was a way of anonymously encouraging other people to check his results while keeping his name out in case he looked like an idiot. Or maybe he had it in for Richard Chevenix and set the whole thing up as an ingenious mind-game to make Chevenix himself look like the idiot.

47 *Silver (Ag)* Known to the Ancients

48 *Cadmium (Cd)* Impurity

During the early nineteenth century in Germany, zinc compounds were being sold as dietary supplements. Unfortunately most of the zinc being used came from the Belgian town of Kelmis where there was an impurity in the rock: cadmium, one of the most toxic elements on the periodic table.

When people started getting sick, the physician Johann Roloff became worried they were eating something poisonous and called on Friedrich Stromeyer, Inspector General of apothecaries, to investigate the supplements. Along with Karl Hermann, owner of the zinc mine, Roloff and Stromeyer eventually identified the offending element in 1817 and named it after *cadmia*, the Latin for zinc, the element it was supposed to be.[48]

49 *Indium (In)* Spectroscopy surprise

Discovered in 1836 by the German chemist Ferdinand Reich by analysing pyrite minerals. Reich was colour-blind and had to hire a man with the outrageous name of Hieronymus Richter to help him identify the colours. By inspecting the minerals, he encountered (or at least was told he encountered) a dark blue beam signifying a new element which he named after indigo, the colour invented by Isaac Newton.[49]

50 Tin (Sn) Known to the Ancients

51 Antimony (Sb) Known to the Ancients

52 Tellurium (Te) Impurity

Discovered in 1782 by Franz-Joseph Müller von Reichtenstein who found it as an impurity in ore obtained from a Hungarian gold mine. He named it from the Greek *tellus*, meaning 'of the Earth'.[50]

53 Iodine (I) Total accident

Discovered in 1811 by the French chemist Bernard Courtois, a saltpetre manufacturer. This was a lucrative business in the 1800s because saltpetre is one of the key ingredients in gunpowder.

The main way of making it was to extract it from burning seaweed and then rinse the ashes in water. The leftover ashy-sludge was typically useless, and the standard procedure was to dissolve it away carefully with sulfuric acid. On one occasion Courtois had finished with a batch, but as he poured out the acid his hand slipped and he added too much. The ash reacted vigorously and spat out a dark purple gas which he christened iodine from the Greek *iodes*, meaning 'violet'.[51]

The remarkable thing is that iodine was apparently being produced every time anyone treated seaweed ash with sulfuric acid. It's just that they usually did it in small amounts so the purple iodine was too sparse to be seen by the naked eye. It was only because Courtois overdid it that he noticed the gas at all.

54 Xenon (Xe) Total accident

Discovered as part of William Ramsay's puddle. Gets its name from the Greek *xenos* meaning 'stranger'.[52]

55 Caesium (Cs) Spectroscopy surprise

Discovered in 1861 by Robert Bunsen using his spectroscopy technique (the first element to be discovered this way) on a sample of mineral water from Durkheim in Germany. The emission beams were bright blue which gave it its name, from the Latin *caesius*, meaning 'sky-blue'.[53]

56 Barium (Ba) Breakdown surprise

Humphry Davy and his battery strikes for the last time in 1808, on a mineral called baryta, from the Latin *barys*, meaning 'heavy'.[54] Davy holds the record for discovering the greatest number of naturally occurring elements on the periodic table, with a whopping total of seven.

57 lanthanum (La) Breakdown surprise

Discovered in 1843 by Swedish chemist Carl Mosander (Berzelius's house guest) from analysing the mineral cerite. He had already extracted another element from it (cerium) but it took much longer to isolate this one, so he named it from the Greek *ianthanin*, meaning 'to lie in secret'.[55]

58 Cerium (Ce) Breakdown surprise

Discovered in 1825 by Carl Mosander from analysing the same mineral cerite, after which he named it. The mineral itself got its name from Berzelius who named it for the dwarf planet Ceres discovered two years previously.[56]

59 Praseodymium (Pr) Breakdown surprise

When Mosander broke down his cerite mineral in 1825 he extracted the cerium but found another substance left over, which he believed to be a combination of two elements. In 1843 he successfully split this in two. One was the aforementioned element lanthanum and the other was named didymium from the Greek *didymo*, meaning 'twin', since it seemed to have twin properties with lanthanum.

However, Mosander was mistaken. Didymium was actually another composite of two elements – praseodymium and neodymium – which were both isolated in 1885 by Carl Welsbach. He named praseodymium from the Greek *prasios*, meaning 'leek green', since praseodymium formed salts with green colours.[57]

60 Neodymium (Nd) Breakdown surprise

Discovered at the same time as praseodymium by Carl Welsbach in 1885. He named it from the Greek *neos didymos*, meaning 'new twin'.[58]

61 Promethium (Pm) Predicted

Promethium's story is similar to that of technetium in that it is extremely unstable (for reasons of nuclear weirdness). Elements 60 and 62 were known, so finding element 61 was just a matter of time. It was discovered by three American nuclear physicists – Jacob Marinsky, Lawrence Glendenin and Charles Coryell – in 1945 while researching uranium for the war effort.

The three discovered traces of an element with an unrecorded half-life that they realised was element 61. It gets its name from the Greek god Prometheus who was punished for stealing fire and giving it to humans – a suggestion made by Charles Coryell's wife, Grace, who wanted a name that would reflect both the power and danger of nuclear energy.[59]

62 Samarium (Sm) Spectroscopy surprise

Discovered in 1879 by Paul-Émile Lecoq using spectroscopy on a mineral called samarskite. Samarskite itself gets its name from the Russian mining official Vasili Samarsky-Bykhovets who supervised the mines where it was found.[60]

63 Europium (Eu) Impurity

Discovered in 1901 by Eugène-Anatole Demarçay while doing work on samarium. The sample of samarium he was using for spectroscopy contained an impurity which he identified as an element.[61] He named it after Europe, itself named for the Greek princess who was kidnapped by Zeus in the form of a bull.

64 Gadolinium (Gd) Spectroscopy surprise

Discovered in 1880 by the Swiss chemist Jean Marignac after spectrosco-pising (definitely a word) the mineral gadolinite, which was named after the Finnish geologist Johan Gadolin – one of Marignac's heroes. Gadolin had proposed the possible existence of this element from a piece of rock he found in the Ytterby mine (the one which also gave us scandium and yttrium), but said it would be a pity if such an element really existed because there were already too many.[62]

65 Terbium (Tb) Impurity

Discovered by Carl Mosander in 1843 by analysing a piece of yttria and noticing an unwelcome contamination. He *also* named his discovery after the village mine of Ytterby in Sweden, the same place that gave us yttrium, scandium and gadolinium.[63]

66 Dysprosium (Dy) Impurity

Discovered by Paul-Émile Lecoq in 1886 on the mantelpiece of his fire-place. He too was researching yttria and found an impurity which he doggedly pursued for thirty years before finally isolating it. He named it from the Greek *dysprositos*, which means 'hard to get'.[64] And after all that effort, what is dysprosium used for? Nothing. It is the most pointless element on the periodic table.

67 Holmium (Ho) Spectroscopy surprise

Discovered in 1878 by the Swiss chemist Marc Delafontaine while studying the spectroscopic light emissions of, once again, impure yttria.[65] It gets its name from the Latin name for Stockholm, *Holmia*.

68 Erbium (Er) Impurity

Discovered in 1843 by Carl Mosander as an impurity in a sample of, yup, you guessed it, yttria. Not exactly the most creative one for names, Mosander named it in exactly the same way he named his earlier discovery of terbium – after the Swedish village of Ytterby.[66]

69 Thulium (Tm) Impurity

Discovered in 1879 by the Swedish chemist Per Cleve. This time it was found as an impurity in a sample of a different mineral called erbia . . . which also came from Ytterby. He named it from *Thule*, the Greek word for Scandinavia.[67]

70 Ytterbium (Yb) Impurity

Isolated by Carl Welsbach in 1907 from impure samples of gadolinite.[68] Guess which Swedish village inspired the name . . .

71 Lutetium (Lu) Impurity

Isolated by the French chemist Georges Urbain in 1907 from impure samples of ytterbia – a mineral found commonly in a certain settlement on the Stockholm archipelago. He named it, thankfully, after his home city of Paris, which in Latin is *Lutetia*.[69]

72 *Hafnium (Hf)* Predicted

Predicted by Mendeleev's table, hafnium was discovered in a sample of Norwegian zircon in 1911 by Dutch physicist Dirk Coster and Hungarian radiochemist George de Hevesy. It had originally been claimed by Georges Urbain, until his sample turned out to be more of his earlier discovery lutetium, just in purer form.[70] Coster and de Hevesy named the element from the Latin name for Copenhagen, *Hafnia*.

73 *Tantalum (Ta)* Breakdown surprise

Discovered by the Swedish chemist Anders Ekeberg in 1802 by breaking down samples of a mineral curiously named yttria.[71] He named it after the Greek mythological character Tantalus who, wanting to test the gods' ability to know everything, logically murdered his son, chopped him up and offered him to the gods in a pie.

As punishment for doing this, Tantalus was imprisoned in a room with water that would drain away any time he bent to drink it (hence the word tantalising). Ekeberg named the element after this gruesome tale because tantalum didn't absorb water either . . . obviously the first thing you think of when you read that story.

74 *Tungsten (W)* Breakdown surprise

Discovered in 1783 by the Spanish brothers Fausto and Juan José Elhuyar by breaking down the mineral wolframite.[72] The name of the mineral comes from the German *wolf rahm*, which means 'wolf's foam' – nobody knows how the mineral got this name originally. In Germany the element is still called wolfram after the mineral, but in most English-speaking countries it is called tungsten, the Swedish for 'heavy rock'.

75 Rhenium (Re) Breakdown surprise

Discovered in 1925 by the German chemists Walter and Ida Noddack from breaking down the minerals columbite, gadolinite and molybdenite. They named it for the River Rhine (Rhein) in Germany.[73]

76 Osmium (Os) Impurity

Discovered in 1803 by the English scientist Smithson Tennant. Whenever platinum was dissolved in aqua regia acid, there was always an impurity in the solution afterwards which wouldn't dissolve. Platinum apparently always came mixed with this undesirable greyish powder which had previously been mistaken for graphite. Nobody before Tennant seems to have thought to check it out. He analysed it and extracted a new element with a distinct smoky smell, which he named from the Greek *osme*, meaning 'smell'.[74]

77 Iridium (Ir) Impurity

Discovered at the same time as osmium by Smithson Tennant as another impurity found in platinum ore. This element formed beautifully coloured salts so he named it for the Greek goddess of the rainbow, Iris.[75]

78 Platinum (Pt) Known to the Ancients

79 Gold (Au) Known to the Ancients

80 Mercury (Hg) Known to the Ancients

81 Thallium (Tl) Impurity

Remember how selenium was discovered during the manufacture of sulfuric acid because the pyrite used to obtain the sulfur contained small amounts of selenium impurity and it was mucking up the reaction chamber? No?

Well, thallium was discovered – independently – in exactly the same way. The pyrite which yielded selenium in 1817 came from Gripsholm in Sweden, while the pyrite in this story came from the Harz mountains in 1861.

Once again, pyrite was cooked in air and dissolved in water to make sulfuric acid, but once again a powdery build-up gunked up the chamber, ruining the process and shutting down production.

The owners of the factory called on the services of English scientist William Crookes to investigate, and he decided to subject the powder to spectroscopy. He discovered it produced undocumented bright green emissions that looked similar to plant shoots so he named it from the Greek *thallos*, meaning 'twig'.

Crookes published his discovery, but the following year the French scientist Claude-August Lamy discovered thallium independently. Lamy was awarded a medal by the Royal Society in London which prompted Crookes to write in angrily, pointing out that he had done it first and should get a medal too. In the end, both scientists were given copy-cat medals for the copy-cat discovery of an element that was copy-catting selenium.[76]

82 Lead (Pb) Known to the Ancients

83 Bismuth (Bi) Unknowingly known to the Ancients

Bismuth was technically known in the ancient world but it was mistaken at different times for lead, tin, antimony and zinc. It had properties that lay somewhere in between all those metals, so whenever anyone was working with it they simply assumed it was an impure form of one of the others. Bismuth was therefore mistaken for an impurity when in fact it was a pure element.

The sixteenth-century Swiss physician Paracelsus came closest to realising it was something distinct when he called it a 'bastard half-metal', but he wrongly concluded it was nature failing to create lead, tin, antimony or zinc.

The realisation that bismuth was an element came in 1753 from the French chemist Claude Geoffrey. Geoffrey took pieces of the other four metals alongside bismuth and subjected them to identical reactions. He showed that bismuth's reaction profile could not be explained as being closest to any one of them alone. It had to be a fifth metal in its own right.[77]

84 Polonium (Po) Breakdown surprise

Discovered by Marie and Pierre Curie in 1898 as the result of breaking down pitchblende. They were originally looking to build stocks of uranium (see Part Three), but stumbled across this element in the process. They named it for Marie's home country, Poland.[78]

85 Astatine (At) ?

With astatine we come to a problem because, in a sense, it's never been discovered. Mendeleev's periodic table predicts its existence, but unfortunately it's the rarest element on Earth. It's estimated that no more than 25g exists in the entire planetary crust, which means nobody has ever been able to gather a lump of it big enough to claim they have 'discovered' it.

The closest we can get is to manufacture it in atomic quantities and look for its half-life. So, although a few atoms of astatine were created in 1940 by chemists Dale Corson, Ken MacKenzie and Emilio Segrè at the University of California (who named it from the Greek *astatos*, meaning 'unstable'),[79] it remains un-isolated to this day.

86 Radon (Rn) Breakdown surprise/Spectroscopy surprise

Between 1899 and 1900 three different researchers recorded that the element radium produced tiny quantities of radioactive gas with its own unique spectroscopic signature.

Marie Curie recorded the numbers and simply assumed they were due to a different version of radium. Ernest Rutherford recorded the same data and suggested it might be something other than radium (but he didn't know what) and Friedrich Dorn isolated the gas and suggested it was an element. Therefore all three could be credited as having discovered it, depending on how you define 'discovered'.

Each team had actually stumbled across a different version of the same element (same number of protons but different numbers of neutrons) and given each one its own name. Curie wanted to call it radion, Rutherford wanted to call it emanation and Dorn wanted to call it radon. In 1931 Marie Curie switched to Rutherford's team and endorsed emanation, but by then radon was already catching on, probably because it was simpler.

In 1957 the International Union of Pure and Applied Chemistry ruled that Dorn's proposed name of radon had become the most common and would therefore be the official moniker henceforth. Dorn had called it radon after the element radium itself, just to make things even more confusing.[80]

87 *Francium (Fr)* Impurity

Much like astatine, francium is insanely rare and large quantities have never been isolated. However, in 1939 physicist Marguerite Perey noticed something with her actinium samples. Actinium is a radioactive element which usually comes with a lot of impurities. Perey had gone over all the standard methods to purify it, but found her sample was still not giving out emissions characteristic of pure actinium. She realised there was another element inside, which she named for her home country, France.[81]

88 *Radium (Ra)* Breakdown surprise

Discovered by Marie and Pierre Curie six months after their discovery of polonium in 1898, again by breaking down pitchblende. They named this one after the fact that it was radioactive.[82]

89 *Actinium (Ac)* Breakdown surprise

After Marie and Pierre Curie were done with their pitchblende, they decided to stop for some reason. They handed their leftover samples to their friend André-Louis Debierne who thought he might as well have a look to see if there was another element they had missed. There was. He discovered it the following year and named it from the Greek *aktinos*, which means 'ray'.[83]

90 Thorium (Th) Breakdown surprise

In 1828 the Norwegian priest Morten Esmark was wandering around the island of Lovoya when he happened to notice a black rock he hadn't seen before. He sent it to Jöns Jacob Berzelius for analysis, who broke it down and discovered an element inside,[84] which he named after his favourite Marvel superhero.[85]

91 Protactinium (Pa) Blind reaction

Sort of discovered by William Crookes in 1900 as an impurity in a sample of uranium he was reacting with nitric acid and dissolving in ether. He wasn't able to prove it was a pure element, though. This was achieved in 1918 by Lise Meitner and Otto Hahn (who originally codenamed it 'abracadabra'). Because this element would decay into element 89, actinium, they named it '*protos* actinium' which is Greek for 'the one before actinium', which ultimately became protactinium.[86]

92 Uranium (U) Blind reaction

Discovered in 1789 by Martin Klaproth. He was also mucking about with pitchblende but decided to dissolve it in nitric acid. It yielded a huge amount of a yellow cake-like substance, which he reacted with charcoal and obtained a black powder, which he named after the recently discovered planet that immature science authors like to inspect closely.[87]

Elements 93–118

Of the remaining elements on the periodic table, all but two were created deliberately by taking heavy elements and firing smaller particles at them in the hope of moving further along the periodic table.

The exceptions were elements 99 and 100 (einsteinium and fermium) which were made by accident on 1 November 1952 as by-products of the first thermonuclear explosion. Both were detected in the atmosphere by the American nuclear scientist Albert Ghiorso who was banned from telling anyone due to security concerns, making him the last person in history to truly discover an element.[88]

How rainbows happen

....

When I was a physics teacher, I would always leave my lecture on 'How rainbows happen' to the end of the optical physics course. This was partly because it was a nice real-world application to finish on, but also because it draws on so many different principles it's tricky to explain without going over them in detail first.

Normally I'd do my rainbow lecture with the aid of demonstrations and animations which I don't have the luxury of in a book, but nevertheless here's a brief attempt to explain Isaac Newton's full, technicolour brilliance.

Molecules at the back of our eyes respond to a range of light wavelengths from 700 nanometres (red) to 380 nanometres (violet). Visible light can take any value between those extremes, so in a sense there's an infinite number of colours. It's just that culturally we group them into red, orange, yellow, green, blue, indigo and violet.

Our eyes only have three main types of colour-detecting cells, though. One type responds to long red/orange wavelengths, one to medium yellow/green and one to short blue/violet wavelengths. But suppose our eye receives more than one beam at the same time? If we shine a red and green light into our eyes we trigger both the long and medium cells, which gets averaged out as the middle colour.

Hence red and green light gets perceived by our brain as yellow. That means there's really two types of 'yellow light' in the world: pure yellow, which is about 570 nanometres, but also red + green light overlapped. The same thing happens if we experience medium and short wavelength light together, except this time it's green + blue light = cyan.

Things get difficult if we trigger the long and short wavelength receptors at the same time, i.e. red and blue light. Our brain gets a confusing signal telling us the light is both long and short, so our mind invents a colour to make sense of it. The colour is called 'magenta' or 'pink' and it doesn't have a wavelength of its own. No object in the Universe gives out pink light: it's just an illusion created inside our heads. A pink object is actually giving off red and blue light, but our brains can't handle that.

If an object is giving out red, orange, yellow, green, blue, indigo and violet together, our brain has to do something similar and invent another colour to deal with the overload. This is where it invents 'white' – another non-spectral colour that doesn't exist in the outside world, only in our heads.

The analogy I sometimes use is to imagine a classroom full of children talking at the same time. Your brain isn't capable of picking apart each conversation so it blurs them together into the background hum of a crowd. Our eyes do the same with colour to create white. Black would be equivalent to total silence, i.e. an object not giving out any light, forcing our brains to come up with a shade to explain a colour-gap. That's the first weird bit ... pink, black and white aren't real, they're just what happens in our minds when something is multi-coloured.

Now for the colour of the sun. The sun is white. It only looks yellow sometimes because of an optical illusion. When we look at the sun through a thick layer of sky the different colours of light bounce off atoms in the air at differing degrees. The shorter wavelengths like blue and violet have the most energy so they go ricocheting out across the dome of the sky above us, making it appear green/blue/violet. But because we don't have many violet-sensitive cells in our eyes we miss them and only see the sky as green + blue = cyan.

The light blue you're seeing on a sunny day is actually sunlight that has been thrown off course to a different part of the sky. You're just not seeing all the violet as well. Lower energy colours like red, orange and yellow don't get bounced around in the air so much and stay in a straight line. So, if we look near the sun we see the area around it as yellow/orange and the sky further away as cyan.

At sunrise and sunset we're looking at the sun through a much thicker angle of atmosphere so the effect is more pronounced and the sky around the sun starts glowing orange/yellow, while the sky further out becomes violet enough for us to see it.

In fact, getting weirder, the colour our sun emits more than any other is actually green. We just don't see it because it's also giving off red and blue as well, and our eyes get confused by all the information. Our sun is more green than anything else, which is why plants are green. They're reflecting (rejecting) the most prominent colour because absorbing it would damage them. On a planet orbiting a red giant, the plants would be red.

So there you go: the sun is green and the sky is purple. We just have rubbish brains.

So what happens when a beam of 'white' light enters a raindrop? It all depends on the angle. When a beam of light moves from air into water the overall rate at which the beam moves slows down due to increased density. If the beam hits the boundary between the air and the water at a perfect right angle nothing special happens, it just slows down. But if it hits it slightly askew, the beam will twist as it crosses the boundary.

Think of a car driving from a smooth road onto a patch of sand. If it drives head on from tarmac to sand all that happens is a reduction in speed. But if the car drives onto the sand at an angle, one wheel is slowing before the other, causing the car to turn. Beams of light work in a similar way. If one side of a beam is slowing down as it enters the water, the overall trajectory of the beam will turn, causing a change in direction.

The degree to which the beam changes is related to wavelength. Violet light is bent by a much higher angle than red light (it's more energetic so it ricochets at a greater angle). This means that if you shine a beam of white (multi-coloured) light through a droplet, the different component beams will bend by differing amounts, with red the least and violet the most.

The next step is to understand what happens when a beam of light tries to go from a more dense to a less dense material. We know that beams bend inwards as they go from less dense to more dense mediums, so logically we can see that they do the opposite when trying to re-emerge: they bend outwards.

But if the angle they hit the boundary at is too steep, they bend backwards so much that they actually reflect. Shining a beam of white light on a raindrop will therefore cause the light to bounce back out in the direction of the sun (i.e. you need to have your back to the sun to see the beam of light coming towards you).

If you do look at the sun directly during a rainstorm (don't) you can often see a rainbow, but the rainbows we can safely observe are the ones which are bouncing off the *backs* of the raindrops.

This creates a circular pattern of split light coming towards us from each droplet, with a red ring on the outside and a violet ring on the inside. Every droplet in a rainstorm is making this happen, but the angle you have to be standing at to see it means the ground gets in the way and cuts off the bottom. Rainbows are not arch-shaped at all, they're circles, but you can only see the top half.

Because there are millions of raindrops in a typical rainstorm, each one is sending out a rainbow circle, but since you're only standing in one place you can't see them all. The rainbow you see is thus composed of light from multiple raindrops.

One of the raindrops particularly high in the sky will be sending out a circle of light but your eye is only in position to see the very top of it, so you see a red beam from that drop. Another raindrop slightly to the left is also sending out a circle of light but you're at the wrong angle, so you see the sky behind it.

Below the red drop, another drop is sending out its own circle – but you're in the wrong position to see red this time. Instead, you're in the right place to see orange. Every droplet is emitting a full rainbow but you see the red from one, the orange from another, the yellow from another and so on.

The rainbow you're looking at is therefore unique to you. A person standing next to you is seeing a slightly different rainbow from a slightly different set of drops. You are the only person in the Universe seeing the exact rainbow you're seeing, and if you step left or right, you're seeing different rainbows in each location.

Rainbows are always the same distance and angle away from you, and they move perfectly in sync with your eyeline. This is why you can never get any closer to a rainbow to get the pot of gold – as you take a step forward, you see a new rainbow from the droplets further back.

Double rainbows occur because there are multiple angles at which light can bounce off the back of a drop and produce the effect. Technically, every rainbow you see is a double rainbow, it's just that the second one is usually too faint. If you *do* manage to see the second rainbow, you'll notice that the colours are inverted, with violet at the top and red at the bottom – a result of the light bouncing out of the drop at a steeper angle.

The silver screen

Silver particles and nitrate particles floating in solution are separated from each other and have electrical charges. These charged versions of atoms are called ions, and silver ions are positively charged while nitrate ions are negative.

This balance of charge attraction is what keeps the solution stable. Everything is attracted to everything else, but not strongly enough to overcome the water particles in between them and bond to a solid.

When high-energy light hits the solution, however, it gives energy to the negatively charged nitrate ions. The nitrate ions have negative charge because they have a bonus electron in their structure, but when light hits them, the electrons gain enough energy to break away.

The most appealing place these high-energy electrons will want to go is the positively charged silver ions nearby. The positively charged silver ions pick up the negatively charged electrons and become neutral.

The more silver atoms you get to do this, the more you build up neutral silver. Neutral silver atoms merge nicely with other silver atoms, which leads to the formation of clumps of solid silver which are a beautiful dirt colour.

Most powdered metals are not shiny in the way you might expect. The shine of a metal's surface is a bulk-property due to having lots of electrons in the surface which can move freely. In a small speck, the same thing is going on but not enough to create the laminated sheen we're used to.

Unstable atoms

Protons in the nucleus of an atom have a positive charge, which means they don't like sitting near each other. It would be impossible to make an atomic nucleus from just two protons – it would be like trying to force the ends of two magnets together. This is what neutrons are useful for.

Neutrons are neutrally charged (as the name suggests) so they don't have an impact on charge attraction or repulsion. But there's another force in the nucleus a lot stronger than charge. This strong force is called . . . The Strong Force (clap, clap, physicists).

The Strong Force is a short-range force which acts on neutrons and protons, gluing them together. At a distance, two protons are going to stay away from each other because the charge repulsion will dominate, but if you bring the protons close together with a neutron in between, The Strong Force attraction between the three will be enough to hold them.

As a general rule, you always need an equal or greater number of neutrons than protons in a nucleus to keep it stable and thus, as atoms go up in proton number, we see neutron numbers going up as well.

But protons and neutrons aren't stationary inside the nucleus. They're shuffling and juggling constantly. The more protons and neutrons you have, the more likely you are to get unfavourable arrangements where two protons are pushed close to each other, sometimes disrupting the whole nucleus as they try to repel. That's the bit we understand.

The bit nobody understands is that, for some reason, whenever these radioactive breakdowns occur they always do so in the same way. What gets ejected from a nucleus is either a single neutron or an alpha particle – two protons and two neutrons bound together.

Quantum physics summarised in a stupidly short appendix

Seriously, I wrote a whole book about this topic. But I suppose – if you don't want to spend any more money because you've already bought this book, *Elemental, Astronomical* and *Brain Detective* for the kids – I'd better give you a brief summary.

It seems like a contradiction that we talk about light being sometimes made from particles and sometimes being made from ripples in the electromagnetic field. This is known as the 'wave particle duality' problem, which has a rather cool and weird solution.

Light is definitely made from particles called photons. But these photons aren't objects in the way we typically imagine them. They're not little pellets of matter flying around in straight lines. They're particles with 'probabilistic properties'.

Photons have properties which define their characteristics, but most of these properties do not get fixed in place until they interact with something. Photons are therefore not really particles until you observe them. Before being observed they exist in a hazy sort of ghost-like state called a 'superposition'.

This sounds strange, because in everyday experience properties of an object don't fluctuate. A green apple is always a green apple no matter what it's interacting with. But photons aren't like that – they have properties which are in flux, only crystallising into something constant when they interact with another particle.

One of these fluctuating properties is, rather unhelpfully for our minds, location. A photon doesn't actually have a fixed place in space until it interacts with something, like a detector or our eye.

A free-floating photon is an abstract thing whose position ripples through the Universe, i.e. it moves as if it were a wave. So while the photon is a particle when we pin it down, when it's travelling freely its location is wave-like. This means light has the apparent quality of moving in waves when not observed, but acting like particles when directly observed.

So what are photons themselves made of? Well, they're made of 'excitations' in the electromagnetic (EM) field. The field all around us is usually set to a value of zero, like a calm lake, but when something agitates the field it creates whirls and eddies that move through it at high speed – photon particles. They're like bumps and knots in the otherwise smooth fabric of the EM field.

A beam of light can therefore also be treated as a wave in the EM field because that's sort of what it is. Light is a collection of energetic fluctuations in the EM field, whose locations ripple through reality in waves.

Depending on the experiment we conduct we sometimes see the wave-like behaviour of the beam, but at other times we see the particle-like behaviour. So is light a particle or a wave? In a very real sense, it's both.

It's not just photons which are like this, either. It turns out *all* particles are wave-like when left to their own devices. All particles are blips in background fields and have locations which vibrate, giving the appearance of waves. You are made up of self-contradictory energy nuggets that can't make their mind up where they even are.

Antiparallel DNA

In DNA's backbone there are alternating phosphate and deoxyribose molecules. The deoxyribose are the important ones for this discussion, and they're pentagonal rings with an oxygen atom at the tip. There's a bunch of other chemical shrubbery hanging off each of the points, but the pentagonal shape is the crucial part:

Deoxyribose

The four chemical bases, A, T, G and C, bond to the deoxyribose pentagons and form the rungs of DNA, but this means they can stack in one of two ways. They can either be arranged with all the oxygen tips pointing the same way, called parallel pairing . . .

. . . or, as Crick realised from looking at Rosalind Franklin's unfairly obtained data, the two backbones can face the opposite way from each other. One of the backbones has all its oxygens pointing up the spiral, while the other has all its oxygens pointing down the spiral. Antiparallel pairing:

Acknowledgements

I'd like to thank my editor Emma Smith for coming up with the idea for this book and my fearless agent Jen Christie for making it happen.

Thank yous also go to Amanda Keats, Sue Viccars and everyone else at Robinson who did edits in order to make the book readable.

Thank you to Simon Snowden, Lucy Wilkinson and Hercules A. Pettitt for giving me suggestions on stories to include.

Thank you to Dave Evans for getting me into science, John Miller for persuading me to teach it and Seishi Shimizu (to whom the book is dedicated) for showing me what my brain could do.

Thank you to Nick Hopkins for encouraging my writing, listening to my endless anecdotes and laughing at my jokes.

Thank you to Ellie for all the hearty intellectual debates and thank you to Jo for tolerating me during tuition sessions.

Thank you to Bree for believing in me, especially when I don't believe in myself.

Thank you to my father, Paul, for always being in my corner . . . and for being OK with me suddenly moving to America!

Finally, thank you to Obi for keeping an eye on my dad and keeping him happy.

References

QUOTATIONS

'The most incredible thing about miracles is that they happen.'

G. K. Chesterton, 'The Blue Cross', *The Storyteller* (September 1910) Cassell & Co.

Part One

'If you could kick the person in the pants responsible for most of your trouble, you wouldn't sit for a month.'

T. Roosevelt, *Theodore Roosevelt on Bravery: Lessons from the Most Courageous Leader of the Twentieth Century* (New York: Skyhorse Publishing, 2015)

'If you saw the mountains on my desk, nothing would surprise you!'

A. Einstein, letter to Kurt Grossman, 9 June 1937

Part Two

'He picked up the lemons that Fate had sent him and started a lemonade stand.'

E. Hubbard, 'Obituary for Marshall Wilder', *The Fra: A Journal of Affirmation*, Vol. 14, No. 5 (1915)

'That which does not kill me, can only make me stronger.'

F. Nietzsche, *Twilight of the Idols* (Leipzig: C. G. Naumann, 1889)

Part Three

'*Surprise is the greatest gift which life can grant us.*'
B. L. Pasternak, *The Poetry of Boris Pasternak*
1917–1959 (New York: Putnam, 1959)

'*We've found that the world is very surprising.*'
J. Polkinghorne, 'An Interview with John Polkinghorne'
(Interviewed by P. Fitzgerald) *The Christian Century*
(January 2008) Christian Century Foundation

Part Four

'*The man of talent is a marksman who hits a target others cannot hit – the
man of genius is a marksman who hits a target others cannot see.*'
A. Schopenhauer, *The World As Will And Idea*
Vol. III Trans. R. B. Haldane, J. Kemp, 6th Ed.
(London: Kegan Paul, Trench, Trubner & Co. 1909)

'*When the right idea finally clicks in place . . . one could kick oneself for not
having the idea earlier.*'
F. Crick, *What Mad Pursuit: A Personal View of
Scientific Discovery* (New York: Basic Books, 1988)

PART ONE: CLUMSINESS

1. J. Needham, H. Ping-Yu, *Science and Civilization in China, Volume 5 Part 7*
   (Cambridge: Cambridge University Press, 1986)
2. R. E. Oesper, 'Christian Friedrich Schönbein. Part II. Experimental Labors'
   J. Chem. Educ., Vol. 6 (1929) pp. 677–85
3. B. Bouffard, 'Inventor of the Month – Who Is Edouard Benedictus?'
   Innovate (12 November 2013)
4. Glass Innovation Centre, *Innovations in Glass* (Corning: Corning Museum
   of Glass, 1999)
5. A. Parkes, Patent No. 1313, 11 May 1865 UK Patent Office, *Patents for Inventions
   – Artificial Leather, Floorcloth, Oilcloth, Oilskin and other Waterproof Fabrics*, p. 255

6. H. Heckman, 'Burn After Viewing, or, Fire in the Vaults: Nitrate Decomposition and Combustibility' *The American Archivist*, Vol. 73, No. 2 (2010) pp. 483–506

7. A. Haynes, 'John Walker, Pharmacist and Inventor of the match' *The Pharmaceutical Journal* (2016)

8. B. K. Pierce, *Trials of an Inventor: Life and Discoveries of Charles Goodyear* (New York: Carlton & Porter, 1866)

9. W. C. Geer, *The Reign of Rubber* (New York: The Century Co., 1922)

10. R. Hunter, M. E. Waddell, *Toy Box Leadership: Leadership Lessons from the Toys you Loved as a Child* (Nashville: Thomas Nelson, 2008)

11. J. Tully, *The Devil's Milk: A Social History of Rubber* (New York: NYU Press, 2011)

12. A. Hoffman, *LSD: My Problem Child* (New York: McGraw Hill Book Company, 1980)

13. J. L. Moreno, *et al.*, 'Metabotropic glutamate mGlu2 receptor is necessary for the pharmacological and behavioral effects induced by hallucinogenic 5-HT2A receptor agonists' *Neuroscience Letters*, Vol. 493, No. 3 (2011) pp. 76–9

14. M. A. Lee, B. Shlain, *Acid Dreams: The Complete Social History of LSD: The CIA, The Sixties and Beyond* (New York: Grove Press, 1994)

15. Author Unknown, 'The Inventor of Saccharin' *Scientific American* (17 July 1886, p. 36)

16. S. W. Junod, 'Sugar: A Cautionary Tale' *Update Magazine, The Bimonthly Publication of the FDA Institute* (July–August 2003)

17. 'Molecule of the week: Saccharin, Jul 01 2019' *American Chemical Society* available from: https://www.acs.org/molecule-of-the-week/archive/s/saccharin.html (accessed 21 February 2023)

18. V. S. Packard, *Processed Foods and the Consumer: Additives, Labeling, Standards, and Nutrition* (Minneapolis: UOM Press, 1976)

19. R. H. Mazur, *Discovery of Aspartame*, First chapter in *Aspartame: Physiology and Biochemistry* L. D. Stegink, L. J. Filer, eds. (Boca Raton: CRC Press, 1984)

20. W. Gratzer, *Eurekas and Euphorias: The Oxford Book of Scientific Anecdotes* (Oxford: OUP, 2002)

21. H. L. F. Helmholtz, *On the Sensations of Tone as a Physiological Basis for the Theory of Music* (4th Ed. 1877, trans. A. J. Ellis 1912)

22. C. Mackenzie, *Alexander Graham Bell: The Man Who Contracted Space* (Boston: Houghton Mifflin Company, 1928)

23. S. Shulman, *The Telephone Gambit: Chasing Alexander Bell's Secret* (New York: Norton & Company, 2008)

24. D. J. Albers, G. L. Alexanderson, C. Reid, *More Mathematical People: Contemporary Conversations* (Orlando: Harcourt Brace Jovanovich, 1990)

25. 'Dr Caroline Coats – Short history of the Pacemaker' *Understanding Animal Research* available from: https://vimeo.com/134722418 (accessed 21 February 2023)

26. W. Greatbatch, *The Making of the Pacemaker: Celebrating a Lifesaving Invention* (New York: Prometheus Books, 2000)

27. L. Colebrook, *Biographical Memoirs of Alexander Fleming, 1881–1956* (London: Royal Society Publishing, 1956)

28. A. Fleming, 'The physiological and antiseptic action of flavine (with some observations on the testing of antiseptics)' *The Lancet,* Vol. 190, Issue 4905 (1917)

29. J. Latson, 'How Being a Slob helped Alexander Fleming Discover Penicillin' *Time* (28 September 2015)

30. R. Hare, *The Birth of Penicillin* (London: Allen & Unwin, 1970)

31. K. Berger, 'Alexander Fleming and the Discovery of Penicillin' *Pharmacy Times* (14 March 2019)

32. W. Kingston, 'Streptomycin, Schatz vs Wakman, and the Balance of Credit for Discovery' *Journal of the History of Medicine and Allied Sciences,* Vol. 59, Issue 3 (2004) pp. 441–62

33. D. P. Levine, 'Vancomycin: A History' *Clinical Infectious Diseases,* Vol. 42, Issue 1 (2006) pp. 5–12

34. J. Suszkiw, 'The Enduring Mystery of "Moldy Mary"' *US Department of Agricultural Research Services* available from: https://tellus.ars.usda.gov/stories/articles/the-enduring-mystery-of-moldy-mary/ (accessed 21 February 2023)

35. R. Bud, *Penicillin: Triumph and Tragedy* (Oxford: OUP, 2007)

36. Barry Marshall interviewed by P. Weintraub, 'The Doctor Who Drank Infectious Broth, Gave Himself an Ulcer, and Solved a Medical Mystery' *Discover Magazine* (8 April 2010)

37. B. Marshall, P. C. Adams, 'helicobacter Pylori: A Nobel pursuit?' *Canadian Journal of Gastroenterology,* Vol. 22, No. 11 (2008) pp. 895–6

38. K. C. Atwood, 'Bacteria, Ulcers and Ostracism? H. Pylori and the making of a myth' *Skeptical Inquirer* Vol. 28 (November–December 2004)

39. J. Okafor, 'The History of Tea Bags from Invention Through Popularity' *TRVST* (24 June 2021)

40. H. Markel, *The Kelloggs: The Battling Brothers of Battle Creek* (New York: Knopf Doubleday Publishing Group, 2017)

41. T. Stevenson, *The Sotheby's Wine Encyclopedia 4th Ed.* (New York: DK Publishing, 2007)

Part Two: Misfortunes and Failures

1. G. T. Fechner, *Autobiography* handwritten and recorded as the appendix in M. Heidelberger, *Nature from Within: Gustav Theodor Fechner and His Psychophysical Worldview* trans. C. Klohr (Pittsburgh: University of Pittsburgh Press, 2004)

2. D. A. Leinhard, 'Roger Sperry's Split Brain Experiments (1959–1968)' *Embryo Project Encyclopedia* (27 December 2017)

3. J. M. Harlow, 'Passage of an Iron Rod Through the Head' *The Boston Medical and Surgical Journal*, Vol. 39, No. 20 (1848)

4. Author Unknown, O. S. Fowler & L. N. Fowler eds., 'A Most Remarkable Case' *The American Phrenological Journal and Repository of Science, Literature and General Intelligence*, Vol. 13 (1851) p. 89

5. J. W. Hamilton, 'The Man Whose Head an Iron Rod Passed Is Still Living', *The Medical and Surgical Reporter*, Vol. 5 (17 November 1860) p. 183

6. S. Manjila, *et al.*, 'Understanding Edward Muybridge: historical review of behavioral alterations after a 19th-century head injury and their multifactorial influence on human life and culture' *Neurosurgery Focus*, Vol. 39, No. 1 (2015)

7. S. Corkin, *Permanent Present Tense: The Man With No Memory, And What He Taught The World* (New York: Penguin, 2013)

8. L. R. Squire, *et al.*, 'Description of brain injury in the amnesiac patient N. A. based on magnetic resonance imaging' *Experimental Neurology*, Vol. 105, No. 1 (1989) pp. 23–35

9. M. Frankel, M. Warren, 'How gut bacteria are controlling your brain' *BBC Future Magazine* (23 January 2023)

10. J. F. Cryan, 'More Than a Gut Feeling' – Address at the Annual Conference of the British Psychological Society's Psychobiology Section, *The Psychologist*, January 2019

11. F. Allen, *Secret Formula: The Inside Story of How Coca-Cola Became the Best-Known Brand in the World* (New York: HarperCollins, 1994)

12. I. Osterloh, 'How I Discovered Viagra' *Cosmos* (27 April, 2015)

13. Interview with John LaMattina, researcher on the project for the podcast *Signal* – 16 June 2016, available from: https://www.statnews.com/2016/06/16/tylenol-drugs-signal-podcast/ (accessed 22 February 2023)

14. M. Boolell, *et al.*, 'Sildenafil: an orally active type 5 cyclic GMP-specific phosphodiesterase inhibitor for the treatment of penile erectile dysfunction' *International Journal of Impotence Research*, Vol. 8, Issue 2 (1996) pp. 47–52

15. I. Goldstein, *et al.*, 'Oral Sildenafil in the Treatment of Erectile Dysfunction' *The New England Journal of Medicine*, Vol. 338, No. 20 (1998) pp. 1397–1404

16. L. Klotz, 'How (not) to communicate new scientific information: A memoir of the famous Brindley lecture,' *BJU Int*, Vol. 96, No. 7 (2005) pp. 956–7

17. W. W. Schultz, *et al.*, 'Magnetic resonance imaging of male and female genitals during coitus and female sexual arousal' *BMJ*, Vol. 319, No. 1596 (1999)

18. J. Achan, *et al.*, 'Quinine, an old anti-malarial drug in a modern world: role in the treatment of malaria' *Malaria Journal*, Vol. 10, Article 114 (2011)

19. S. Garfield, *Mauve: How One Man Invented a Colour That Changed the World* (New York: W. W. Norton & Company, 2001)

20. T. F. G. G. Cova, A. A. C. C. Pais, J. S. S. de Melo, 'Reconstructing the historical synthesis of mauvine from Perkin and Caro: procedure and details' *Scientific Reports*, Vol. 7, No. 6806 (2017)

21. C.P. Biggam, 'Knowledge of whelk dyes and pigments in Anglo-Saxon England' *Anglo-Saxon England*, Vol. 35 (2006) pp. 23–55

22. C. Dickens, 'Perkin's Purple' *All The Year Round* (10 September 1859)

23. P. Homem-de-Mello, *et al.*, '4-Design of dyes for energy transformation: From the interaction with biological systems to application in solar cells' *Green Chemistry and Computational Chemistry, Shared Lessons in Sustainability, Advances in Green and Sustainable Chemistry* (2022) pp. 79–114

24. I don't even know how to reference their website properly without risking getting in trouble, but it's here: https://www.velcro.com/original-thinking/the-velcro-brand-trademark-guidelines/(accessed 22 February 2023)

25. Interview with Harry Coover: 'Harry Coover – 2009 National Medal of Technology & Innovation', *NSTMF* available from: https://www.youtube.com/watch?v=u4CLvR-YN2w (accessed 22 February 2023)

26. N. Skillicon, 'The True Story of Post-It Notes and How They Almost Failed' *Idea to Value Newsletter* (20 April 2017)

27. R. B. Seymour, T. Cheng, *History of Polyolefins, 1–7* (Dordrecht: D. Reidel Publishing Company, 1986)

28. M. Lauzon, 'PE: The resin that helped win World War II' *Plastics News* (6 August 2007)

29. Author Unknown, 'Roy J. Punkett' *Science History Institute* (14 December 2017)

30. R. Cole, 'Teflon: 80 Years of Not Sticking To Things' *The Science Museum Blog* (6 April 2018)

31. A. G. Levine, 'The large horn antenna and the discovery of cosmic microwave background' *American Physical Society* (2009)

32. S. Singh, *Big Bang* (London: HarperCollins, 2010)

33. S. Mitton, *Fred Hoyle: A Life in Science* (Cambridge: Cambridge University Press, 2011)

34. R. A. Alpher, R. C. Herman, 'On the relative abundance of the elements', *Physical Review*, Vol. 74, No. 12 (1948) pp. 1737–42
35. R. Lindner, *The Fifty-Minute Hour* (New York: Bantam, 1958)
36. M. Rokeach, *The Three Christs of Ypsilanti: A Psychological Study* (New York: New York Review Books, 1964)

PART THREE: SURPRISES

1. S. Zielinsky, 'When the Soviet Union Chose the Wrong Side on Genetics and Evolution', *Smithsonian Magazine* (1 February 2010)
2. L. A. Dugatkin, 'The silver fox domestication experiment' *Evolution: Education and Outreach*, Vol. 11, No. 16 (2018)
3. AR Androgen Receptor [Homo Sapiens (human)] *National Library of Medicine*, National Centre for Biotechnology Information, Gene ID: 367
4. E. Ramsden, J. Adams, 'Escaping the Laboratory: The Rodent Experiments of John B. Calhoun and their Cultural Influence', *Journal of Social History*, Vol. 42, issue 3 (2009) pp. 761–97
5. B. K. Alexander, *et al.*, 'Effect of early and later colony housing on oral ingestion of morphine in rats', *Pharmacology, Biochemistry and Behavior*, Vol. 14, Issue 4 (1981) pp. 571–6
6. C. Reed, 'Peter Witt Biography', *DrPeterWitt.com* (March 2016) Available from: https://www.drpeterwitt.com/project/peter-witt-biography/ (accessed 9 November 2022)
7. P. Witt, 'Spider Webs and Drugs', *Scientific American* (1 December 1954)
8. *To Tell the Truth* (TV Series) Series 4, Episode 7, dir. Paul Alter, aired 1972, *CBS Daily*
9. H.-P. Rieder, 'Biological Determination of Toxicity of Pathologic Body Fluids III. Examination of urinary extracts of mental patients with the help of the spider web test', *Psychiatry and Neurology*, Vol. 134, no. 6 (1957) pp. 378–396 original article in German, reviewed in M. L. Throne *et al.*, 'A Critical Review of Endogenous Psychotoxins as a Cause of Schizophrenia', *Canadian Psychiatric Association Journal*, Vol. 12, no. 2 (1967) pp. 159–74
10. J. O. Schmidt, M. S. Blum, W. L. Overal, 'Hemolytic Activities of Stinging Insect Venoms' *Insect Biochemistry and Physiology*, Vol. 1, issue 2 (1983) pp. 155–60
11. R. Feltman, 'This scientist rates and describes insect stings as if they were fine wines' *The Washington Post* (17 March 2015)
12. C. Starr, 'A simple pain scale for field comparison of Hymenopteran stings' *Journal of Entomological Science*, Vol. 20, issue 2 (1985) pp. 225–32

13. J. O. Schmidt, *The Sting of the Wild* (Baltimore: The Johns Hopkins University Press, 2016)
14. J. O. Schmidt, 'Pain and lethality induced by insect stings: An exploratory and correlational study' *Toxins*, Vol. 11, no. 7 (2019)
15. M. Piccolino, M. Bresadola, *Shocking Frogs: Galvani, Volta, and the Electric Origins of Neuroscience* (Oxford: OUP, 2013)
16. M. Pilkington, 'Sparks of Life', *Guardian* (6 October 2004)
17. M. Shelley, Introduction to Frankenstein, *Frankenstein: Or, the Modern Prometheus* (15 October 1831)
18. J. Dalton, *Meteorological Observations and Essays* (Manchester: Harrison & Crosfield, 1834)
19. S. Sanctorius, *De Statica Medicina* (1614)
20. A. Bouvard, *Tables Astronomiques* (Paris: Bachelier and Huzard, 1821)
21. J. Uri, '175 Years Ago: Astronomers Discover Neptune, the Eighth Planet' *NASA History* (22 September 2021)
22. L. Grossman, M. McKee, 'Is the LHC Throwing Away Too Much Data?' *New Scientist* (14 March 2012)
23. I. Newton, *A Theory Concerning Light and Colours, 1675* (Archives of Trinity College Cambridge)
24. E. Tretkoff, *et al.*, 'This Month in Physics History – July 1820: Ørsted and Electromagnetism' *American Physical Society News*, Vol. 17, No. 7 (2008)
25. B. Mahon, *The Man Who Changed Everything: The Life of James Clerk Maxwell* (New York: Wiley, 2004)
26. W. Herschel, 'Investigation of the Powers of the Prismatic Colours to Heat and Illuminate Objects; With Remarks, That Prove the Different Refrangibility of Radiant Heat. To Which is Added, an Inquiry into the Method of Viewing the Sun Advantageously, with Telescopes of Large Apertures and High Magnifying Powers' *Philosophical Transactions of the Royal Society*, Vol. 90 (1800) pp. 255–83
27. J. Frercksa, H. Weberb, G. Wiesenfeldt, 'Reception and discovery: the nature of Johann Wilhelm Ritter's invisible rays', *Studies in History and Philosophy of Science A*, Vol. 40, No. 2 (2009) pp. 143–56
28. T. P. Garrett, 'The Wonderful Development of Photography', *The Art World*, Vol. 2, No. 5 (1917) pp. 489–91
29. S. Strickland, 'The Ideology of Self-Knowledge and the Practice of Self-Experimentation' *Eighteenth Century Studies* V. 31, no. 4 (1998) pp. 453–71
30. The Microwave Service Company, *The History & Inventor of the Microwave Oven* (26 March 2012)

31. L. R. Reynolds, 'The History of the Microwave Oven', Lecture given at the 24th Microwave Power Symposium in 1989, transcribed in the *Journal of The International Microwave Power Institute*, Vol. 10, No. 5 (1989)

32. W. R. Nitske, *The Life of Wilhelm Conrad Röntgen: Discoverer of the X-Ray* (Tucson: University of Arizona Press, 1971)

33. T. J. Jorgensen, *Strange Glow: The Story of Radiation* (Princeton: Princeton University Press, 2017)

34. R. W. Chabay, B. A. Sherwood, *Matter and Interactions* (Hoboken: Wiley, 2002)

35. H. C. von Bayer, *Taming the Atom: The Emergence of the Visible Microworld* (New York: Random House, 1992)

36. N. Blaedel, *Harmony and Unity: The Life of Niels Bohr* (Lexington: Plunkett Lane Press, 2017)

37. E. Fermi, *et al.*, 'Radioactivity Caused by Neutron Bombardment', *La Ricerca Scientifica*, Vol. 5, No. 1 (1934) pp. 452–3

38. I. Noddack, 'On Element 93', *Zeitschrift fur Angewandte Chemie*, Vol. 47 (1934) p. 653

39. J. T. Armstrong, 'Technetium: The Element That Was Discovered Twice' *National Institute of Standards and Technology* (16 October 2008)

40. Author Unknown, 'Who Ordered That?', *Nature Editorial*, Vol. 531 (2016) pp. 139–140

41. M. L. Perl, *et al.*, 'Evidence for Anomalous Lepton Production in e+ e- Annihilation', *Physical Review Letters*, Vol. 35, No. 22 (1975)

42. R. P. Feynman, *QED: The Strange Theory of Light and Matter* (London: Penguin, 1985)

Part Four: Eurekas

1. Vitruvius, *The Ten Books on Architecture*, Book 9 (approx. 30BCE)

2. Plutarch, *The Parallel Lives – The Life of Marcellus*, Section 14 (Unknown authorship date, First Century)

3. C. Rorres, H. Harris, 'A Formidable War Machine: Construction and Operation of Archimedes' Iron Hand' *Symposium on Extraordinary Machines and Structures in Antiquity*, Olympia, Greece, 19–24 August (2001) pp. 1–18

4. H. E. Schwarz, 'Super Soaker Inventor – Lonnie Johnson' *STEM trailblazer BIOS* (Minneapolis: Lerner Classroom, 2017) [20]

5. W. Stukeley, *Memoirs of Sir Isaac Newton's Life* (1752) which contains an interview with Isaac Newton in which he relates the story of the apple

6. I. D'Israeli, *Curiosities of Literature Vol. 1* (London: Frederick Warne and Co., 1881)

7. Col T. W. M. Draper, *The Bemis History of Genealogy* (San Francisco: Self-published, 1900)

8. L. Bleiberg, 'The Clock That Changed The Meaning of Time' *BBC Travel Magazine* (7 September 2016)

9. G. Shaw, Percy's granddaughter reflecting on her grandfather's legacy in a brief interview with the BBC titled 'Cat's eyes: How a pub trip made the world's roads safer' (4 February 2023) available from: https://www.bbc.com/news/av/stories-64512319 (accessed 23 February 2023)

10. J. Plester, 'Weatherwatch: Percy Shaw and the invention of the cat's eye reflector' *Guardian* (3 December 2018)

11. This joke is often attributed to Ken Dodd although as with many jokes the original source is unverified. If it was you, then bravo. Seriously. Bravo.

12. O. T. Avery, C. M. Macleod, M. McCarty, 'Studies on the Chemical Nature of the Substance Inducing Transformation of Pneumococcal Types: Induction of Transformation by a Deoxyribonucleic Acid Fraction Isolated from Pneumococcus Type III', *Journal of Experimental Medicine*, Vol. 79, Issue 2 (1944) pp. 137–58

13. A. Sayre, *Rosalind Franklin and DNA* (London: W. W. Norton & Company, 1975)

14. M. Wilkins, *The Third Man of the Double Helix: An Autobiography* (Oxford: OUP, 2003)

15. A. Klug, 'The discovery of the DNA double helix' *Journal of Molecular Biology*, Vol. 335, Issue 1 (2004) pp. 3–26

16. J. D. Watson, *The Double Helix* (New York: Signet, 1968)

17. F. H. C. Crick, J. D. Watson 'Molecular Structure of Nucleic Acids', *Nature*, Vol. 171, No. 4356 (1953)

18. Op. cit. Watson 1968

19. J. H. Richardson, Interview with James Watson – 'James Watson: What I've Learned', *Esquire* (19 October 2007)

20. *American Masters* (TV Series) Series 33, Episode 1 'Decoding Watson', dir. Mark Mannucci, aired 19 December 2018, *PBS*

21. Author Unknown, 'Harpic: Under the Microscope' *Reckitt* (16 December 2020) available from: https://www.reckitt.com/newsroom/latest-news/news/2020/december/harpic-under-the-microscope/ (accessed 23 February 2023)

THE PERIODIC TABLE OF 'HUH?'

1. R. Boyle, *Tracts, Containing New Experiments, touch the Relation betwixt Flame and Air and about Explosions* (London: Richard Davis, 1672)

2. H. Muir, *Eureka: Science's Greatest Thinkers and their Key Breakthroughs* (London: Quercus, 2012)

3. B. B. Nath, *The Story of Helium and the Birth of Astrophysics* (New York: Springer, 2012)

4. M. E. Weeks, *The Discovery of the Elements*, 6th Edition (Easton: Mack Printing Company, 1960)

5. Royal Society of Chemistry, *Elements and Periodic Table History – Beryllium*, RSC Publications (2023)

6. J. L. Marshall, 'Humphry Davy and the Voltaic Pile' *Chem 13 News Magazine* (University of Waterloo: April 2019)

7. Op. cit. Weeks 1960

8. M. I. A. Chaptal, *Elements of Chemistry 3rd Ed Vol. I* (London: G. G. and J. Robinson, 1800)

9. R. Bugaj, 'Michal Sedziwoj – Treatise on the Philosopher's Stone' *Library of Puzzles*, Vol. 164 (1971) pp. 83–4

10. Op. cit. Weeks 1960

11. R. Harre, *Great Scientific Experiments: Twenty Experiments That Changed Our View of the World* (London: Harrap, 1974)

12. R. Toon, 'The discovery of fluorine', *Education in Chemistry* (1 September 2011)

13. D. A. McQuarrie, P. A. Rock, E. B. Gallogly, *General Chemistry 4th Ed.* (New York: University Science Books, 2010)

14. M. S. Muller, 'A pinch of sodium', *Nature*, Vol. 3, No. 974 (2011)

15. E. Katz, 'Electrochemical contributions: Sir Humphry Davy (1778–1829)', *Electrochemical Science Advances* (4 May 2021)

16. T. Geller, 'Aluminum: Common Metal, Uncommon Past' *Distillations*, a publication of the Science History Institute (2 December 2007)

17. Op. cit. Weeks 1960

18. J. Emsley, *The Shocking History of Phosphorus: A Biography of the Devil's Element* (London: Pan Books, 2000)

19. C. W. Scheele, *On Manganese and Its Properties* (1774) collected as the first essay in *The Early History of Chlorine* Eds. unknown (Edinburgh: The Alembic Club, 1912)

20. Op. cit. McQuarrie 2010

21. C. Woodford, *The Elements: Potassium* (London: Cavendish Square, 2003)

22. D. Knight, *Humphry Davy: Science and Power* (Cambridge: Cambridge University Press, 1998)

23. C. T. Horovitz, *Discovery and History*, first chapter in *Scandium: Its Occurrence, Chemistry, Physics, Metallurgy, Biology and Technology*, C. T. Horovitz Ed., (London: Academic Press, 1975)

24. K. L. Housley, *Black Sand: The History of Titanium* (New York: Metal Management Inc., 2007)

25. R. Bowell, *An Introduction to Vanadium: Chemistry, Occurrences and Applications* (New York: Nova Science Publishers, 2019)

26. Royal Society of Chemistry, *Elements and Periodic Table History – Chromium*, RSC Publications (2023)

27. B. Knapp, *Elements: Iron, Chromium and Manganese* (Danbury: Grolier Education, 2002)

28. I. Asimov, *Words of Science* (New York: Signet, 1959)

29. Ibid.

30. Op. cit. Weeks 1960

31. J. A. Bridgman, *Gallium – A Thesis, Presented to the Faculty of the Graduate School of Cornell University for the Degree of Doctor of Philosophy* (New York: Cornell University Press, 1917)

32. M. E. Weeks, 'The discovery of the Elements 13: Some elements predicted by Mendeleeff', *Journal of Chemical Education*, Vol. 9, Issue 9 (1932) pp. 1605–19

33. Op. cit. Weeks 1960

34. R. Boyd, 'Selenium Stories', *Nature Chemistry*, Vol. 3, No. 570 (2011)

35. M. Balard, 'Memoir on a peculiar Substance contained in Sea Water', *Annals of Philosophy*, Vol. 12, Article 13 (1826) pp. 381–7

36. Op. cit. McQuarrie 2010

37. M. Lozinsek, G. J. Schrobilgen, 'The world of krypton revisited', *Nature Chemistry*, Vol. 8, No. 732 (2016)

38. I. Georgescu, 'Rubidium round-the-clock', *Nature Chemistry*, Vol. 7, No. 1034 (2015)

39. F. X. Coudert, 'Strontium's scarlet sparkles', *Nature Chemistry*, Vol. 7, No. 940 (2015)

40. P. Diner, 'Yttrium from Ytterby', *Nature Chemistry*, Vol. 8, No. 192 (2016)

41. J. W. Marden, M. N. Rich, *Investigations of Zirconium, with Especial Reference to the Metal and Oxide, Historical Review and a Bibliography*, Bulletin 186, Mineral Technology 25, Department of the Interior, Washington Government Printing Office (1922)

42. M.A. Tarselli, 'Subtle niobium', *Nature Chemistry*, Vol. 7, No. 180 (2015)

43. Molybdenum history, *The International Molybdenum Association*, available from: https://www.imoa.info/molybdenum/molybdenum-history.php (accessed 24 February 2023)

44. F. A. A. de Jonge, E. K. J. Pauwels, 'Technetium, the missing element', *European Journal of Nuclear Medicine*, Vol. 23 (1996) pp. 336–44
45. V. N. Pitchkov, 'The Discovery of Ruthenium', *Platinum Metals Review*, Vol. 40, No. 4 (1996) p. 181
46. Royal Society of Chemistry, *Elements and Periodic Table History – Rhodium*, RSC Publications (2023)
47. M. C. Usselman, 'The Wollaston/Chevenix controversy over the elemental nature of palladium: A curious episode in the history of chemistry', *Annals of Science*, Vol. 35, Issue 6 (1978) pp. 551–79
48. N. V. Tarakina, B. Verberck, 'A portrait of cadmium', *Nature Chemistry*, Vol. 9, No. 96 (2017)
49. I. Asimov, *Asimov's Biographical Encyclopedia of Science and Technology* (New York: Doubleday, 1964)
50. J. Ibers, 'Tellurium in a twist', *Nature Chemistry*, Vol. 1, No. 508 (2009)
51. Op. cit. Asimov 1959
52. Op. cit. McQuarrie 2010
53. Op. cit. Georgescu 2015
54. Op. cit Knight 1998
55. M. Fontani, M. Costa, M. V. Orna, *The Lost Elements: The Periodic Table's Shadow Side* (Oxford: OUP, 2015)
56. Ibid.
57. Ibid.
58. Ibid.
59. S. Cantrill, 'Promethium puzzles', *Nature Chemistry*, Vol. 10, No. 1270 (2018)
60. Op. cit. Weeks 1960
61. Ibid.
62. P. Pyykko, 'Magically magnetic Gadolinium', *Nature Chemistry*, Vol. 7, No. 680 (2015)
63. L. P. Wilder, *Gadolinium and Terbium: Chemical and Optical Properties, Sources and Applications* (New York: Nova Science Publishers, 2014)
64. J. Emsley, *Nature's Building Blocks: An A–Z Guide to the Elements* (Oxford: OUP, 2001)
65. Ibid.
66. Ibid.
67. P. Arnold, 'Thoroughly enthralling thulium', *Nature Chemistry*, Vol. 9, No. 1288 (2017)
68. Op. cit. Emsley 2001
69. Op. cit. Weeks 1960
70. S. C. Burdette, B. F. Thornton, 'Hafnium the lutecium I used to be', *Nature Chemistry*, Vol. 10, No. 1074 (2018)

71. Op. cit. Weeks 1960
72. O. Sacks, *Uncle Tungsten: Memories of a Chemical Boyhood* (New York: Vintage, 2002)
73. Op. cit. Weeks 1960
74. Op. cit. Emsley 2001
75. Ibid.
76. Royal Society of Chemistry, *Elements and Periodic Table History – Thallium*, RSC Publications (2023)
77. Bismuth entry, *Encyclopedia Britannica, 11th Ed., Vol. 4* 'Bisharin – Calgary' (New York: The Encyclopedia Britannica company, 1910)
78. E. Curie, *Madame Curie: A Biography*, Trans. V. Sheean (Cambridge: Da Capo Press, 2001)
79. D. R. Corson, K. R. MacKenzie, E. Segrè, 'Artificially Radioactive Element 85', *Physical Review,* Vol. 58, No. 672 (1940)
80. B. F. Thornton, S. C. Burdette, 'Recalling radon's recognition' *Nature Chemistry*, Vol. 5, No. 804 (2013)
81. Royal Society of Chemistry, *Elements and Periodic Table History – Francium*, RSC Publications (2023)
82. C. Hobb, H. Goldwhite, *Creations of Fire: Chemistry's Lively History from Alchemy to the Atomic Age* (New York: Basic Books, 1995) NB: This was the second science book I ever read.
83. Ibid.
84. A. J. Ihde, *The Development of Modern Chemistry* (New York: Dover Books, 2012)
85. This book. This book you're reading says that.
86. R. L. Sime, 'The discovery of protactinium', *Journal of Chemical Education*, Vol. 63, Issue 8 (1986) p. 653
87. M. J. Monreal, P. L. Diaconescu, 'The riches of uranium', *Nature Chemistry*, Vol. 2, No. 424 (2010)
88. A. Ghiorso, 'Einsteinium and Fermium', *Chemical Engineering News*, Vol. 81, No. 36 (2003) pp. 174–5

Index

Note: page numbers in **bold** refer to diagrams.